COLDWATER INDIAN ARTIFACTS PRICE GUIDE

Volume I

By

Doug Puckett

ISBN 0-9637157-2-0

First Printing - July 1993
Second Printing - November 1994
Third Printing - June 1998

INTRODUCTION

What tribe made it?
How old is it?
Was the side of that arrowhead chipped so that it would rotate when shot?
How much is it worth?

These questions are the predominant source of discussion with new collectors, school groups, the inquisitive visitor to your home, or the farmer or hunter with a few arrowheads.

Is it a Dalton or a resharpened Clovis?
What other point types were found on that site?
Is it a Lost Lake or a Hardin?
Have you ever seen another one like it?

All you have to do is get two collectors together and these topics or similar ones are guaranteed.

To many of you these questions may seem extremely elementary; to others, you may not have a clue. The tribe question seems to be a fixation with anyone seeing their first flint artifact and is easily attributable to a glich in our educational system that teaches the historic and nothing of the prehistoric. It seems to fall into the hands of collectors to research and preserve the prehistory of the United States and it is a worthy goal.

Indian relic collecting is one of the fastest growing hobbies as can easily be seen by the ever increasing number of visitors and exhibitors at the many relic shows around the country. Many "old time" collectors feel that this rapidly escalating interest is due to an increased knowledge of the age and use of the relics; the family aspect of the hobby in which all members of the family can contribute; and last, but certainly not least, the sky-rocketing value of Indian relics. No other hobby will allow you to take the family to a plowed field, find the object of your search at no cost, find the age and use of your relic, and then apply a value to it. Antique, coin, stamp, gun, etc. collectors can only dream of the no cost part of our hobby. Collecting can get no better than this.

For the promotion of Indian relic collecting, Doug has written this book with the novice as well as the advanced collector in mind. Although point typology is sometimes a mind boggling proposition of best, Doug has established a format to simplify this task. Not only has he traveled extensively throughout the Southeast to photograph the best available examples of known point types, he has also devoted many hours of research, both written and verbal, to the accuracy of this book. The proof of the worth of any collector book is in the range of examples illustrated for each named type and that is what makes this book a must. Where possible, examples from the nearly exhausted to the pristine are pictured. Whether this is your first relic book, or if you have an extensive library, this book will greatly enhance the enjoyment of your relic collecting.

Unfortunately, this must end on a negative note. For those of you that wish to improve your collections by purchase or trade, be warned that there are a few unscrupulous persons waiting to take your money or good relics in exchange for excellent fakes or reconditioned authentic relics. In the past, fakes were fairly easy to spot, but now there are some on the market that will fool the most advanced collector. There cannot be enough emphasis placed on the necessity for studying authentic relics as to the manufacturing techniques and the materials used to produce them. Also, please condition yourself to deal only with reputable dealers and collectors that guarantee their relics. Study, ask questions, discipline, and resist greed to keep your collecting fires burning instead of having them snuffed out by a bad relic at a high price. Remember, "If it's to good to be true, it probably is."

Good hunting.

Dale Strader, President
Green River Archaeological Society of Kentucky

Basic Point Shapes and Features

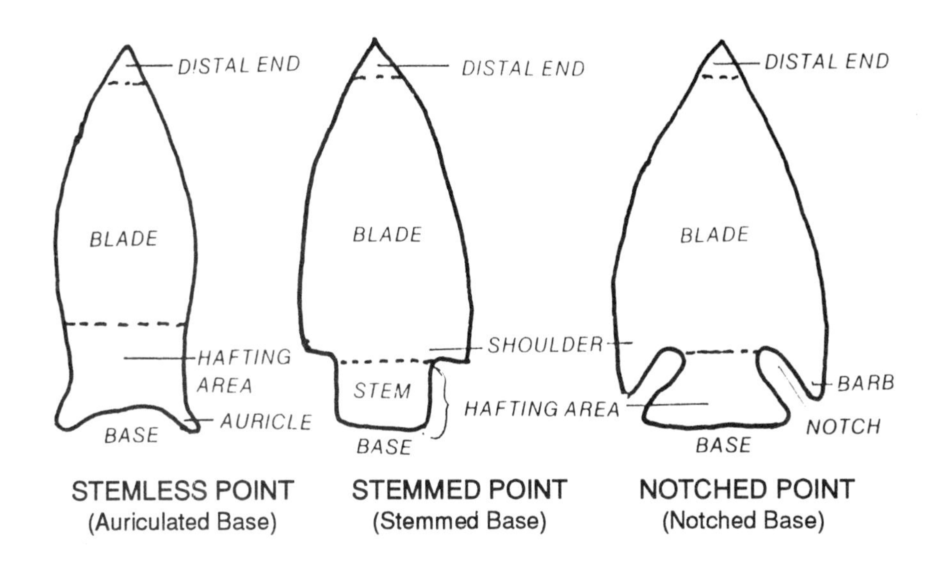

STEMLESS POINT (Auriculated Base) | STEMMED POINT (Stemmed Base) | NOTCHED POINT (Notched Base)

Culture Period

Paleo ....8,000 B.C. and Before

Transitional Paleo Period ....8,000 B.C. to 6,000 B.C.

Early Archaic..............................6,000 B.C. to 3,000 B.C.

Archaic..3,000 B.C. to 1,500 B.C.

Woodland ....1,500 B.C. to 1,000 A.D.

Mississippian.............................1,000 A.D. to 1,600 A.D.

Historic.......................................1,600 A.D. to 1,820 A.D.

DEDICATION

This book is dedicated to the memory of those who have gone before and to the hope that our childrens' children will be able to enjoy artifact collecting and study as we have.

****

Special thanks to John Powell for providing the author with photographs of many of the point types of Florida.

And to Charley Smith for helping with the compilation of this book and to Dale Holley for wearing blisters on his fingers while cutting out photos.

CONTRIBUTORS

A Special thanks to all the members of the Kentucky, Tennessee, Alabama, Georgia, Mississippi and Florida Society who gave of their time to help me locate artifacts and those who allowed me to photograph their collection.

Charley Smith
Wessley Hamilton
Gary Williams
Bimbo Kohen
Walter Farr
John Powell
Barry Waters
Dennis Bushey
David Waite
Rick Weems
Charles Ray
Michael McKey
K. B. McEltree
Bill Mann
Dale Strader
A. W. Beinlich
Dennis Quientavalle
George Stephens
Willis Family

Sid Hite
Ralph Allen
Tom Hendrix
Wallace Culpepper
Indian Mound Museum
Bob Rea
Alice Rea
Horace Holland Collections
Louise Young
Tommy Minor
James D. Stanfield
Howard Cross
Jimmy Wilkes Family
Son Anderson
Troy Futral
Tom Davis
Jack Wilhoite

Webb Giles
James Roshto
Mike McCoy
David Sewell
Bridget Sewell
Stan Griggs
Dale Holley
Russel Cave Museum
Charles Conrades
Vicki Conrades
Ronnie Hawk
Joseph Love
Richard Anderson
Charles Moore
Tim Guise
Doug Puckett
David Smith

Table of Contents

****

****

****

****

****

Abbey

(Pictures 1-A thru 1-C)

Abbey is a medium sized, stemmed point with incurvate blade edges. The blade edges are beveled on both sides of each face.

Sizes range from about 1-1/2 inches to 5 inches, averaging 1-3/4 inches long, with an average width at the shoulder of 1-1/2 inches.

The cross-section of this point is flattened. The shoulders are expanded and most often horizontal but may taper. The distal end is acute. The blade edges are, in rare cases, serrated. The stem ranges from straight to slightly expanded. The basal edge is either straight or slightly expanded and is usually thinned. The shoulders may be pointed, barbed, or somewhat rounded.

The flaking is broad, shallow and random. The blade and stem edges show retouch by shallow, regular, pressure flaking.

The type gets its name from Abbey Creek in Henry and Houston Counties of Alabama. Their association with Elora and Maples seems to indicate Archaic Period use. The point was named by David Hulse.

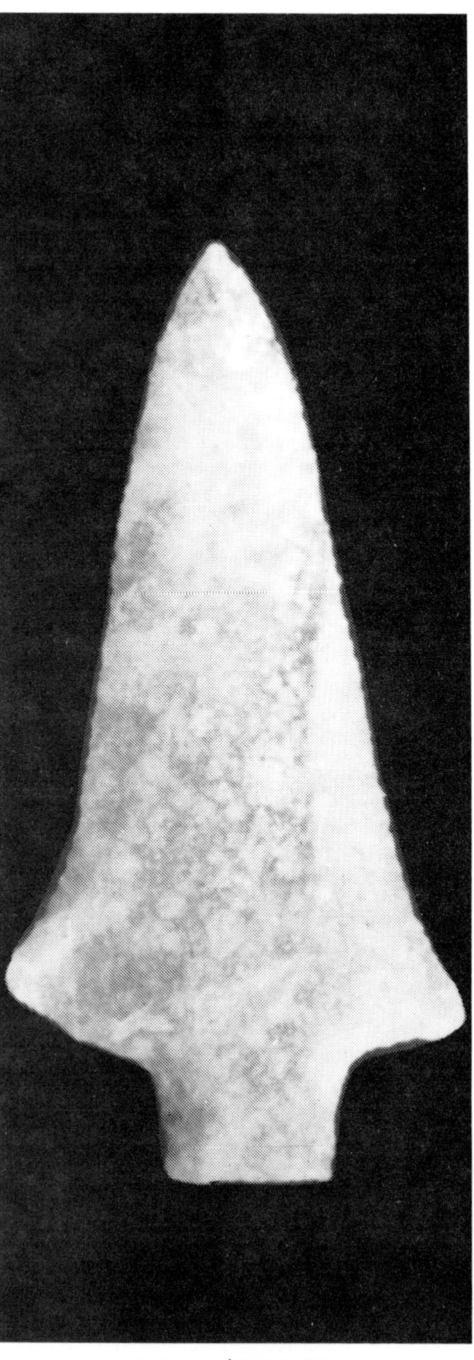

1-A - $300.00

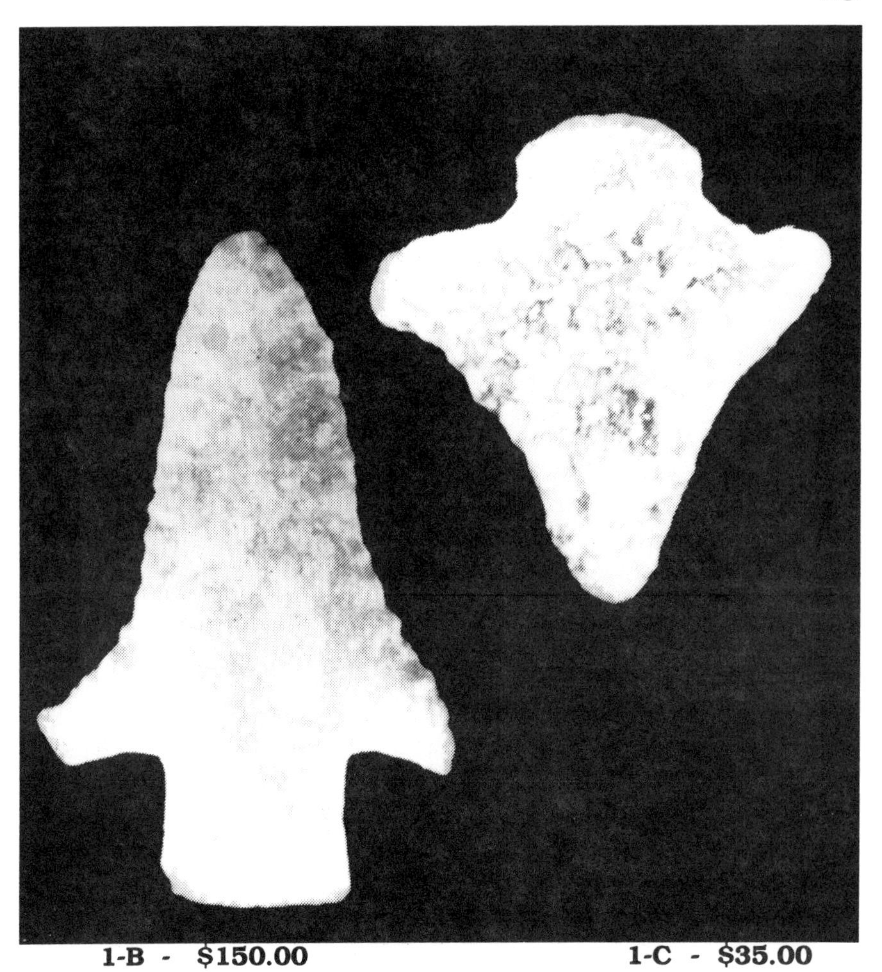

1-B - $150.00 **1-C - $35.00**

Adena *(Pictures 2-A thru 2-C)*

Adena is a medium to large stemmed point. The stem is long and very broad, often rounded. Workmanship ranges from rough to extremely well worked.

These points average around 2 to 2-1/2 inches in length and 1-1/4 to 1-3/8 inches at shoulders.

Adena has a biconvex cross-section. The shoulders may be from 90° angles to the stem to slightly tapered toward the distal end, occasionally weakly barbed. The blade is usually excurvate with an acute distal end. Stems may be straight, contracted or slightly expanded. The stem base is straight or excurvate and is often lightly ground.

(continued on next page)

The stem and blade show strong random flaking and finer retouch along most edges.

2-A - $450.00 - **2-B - $150.00** **2-C - $75.00**

Adena Narrow Stemmed

(Pictures 3-A thru 3-B)

Adena Narrow Stemmed is a medium to large point with a long excurvate blade and a long rounded stem.

The average length is 3 to 4 inches and width averaging 1 inch to 1-1/2 inches at shoulders.

The cross-section is biconvex. The shoulders range from 90° angles to the stem to a slight taper toward the distal end. The stem may be rounded or with straight sides and excurvate basal edge. It should be noted the stem is notably narrower in relation to blade width than the Bell Adena.

(continued on next page)

Blade and stem are shaped by broad percussion flaking showing a rather uniform flake scar pattern. On occasion this pattern may approach collateral flaking. All edges were treated with secondary flaking and fine retouch.

This type, a variant of the Classic Adena, was named by Cambron. Some sites produce both Adena types, where other sites have only one of the two types occurring. Adena appears to be associated with early Woodland - late Archaic cultures and are a wide spread type in the Eastern U.S.

3-A - $500.00 **3-B - $150.00**

Afton

(Picture 4-A)

Afton is a medium to large stemmed point. The blade is straight with angular point.

Average length is about 1-3/4 inches. The average width at shoulder is about 1 inch wide.

Afton has a flattened cross-section. Shoulders are tapered with a straight stem. The blade is parallel angular with apiculate distal end.

The blade and stem are shaped by broad percussion flaking with fine retouch along edges.

The type was named by Bell and Hall for examples found near Afton, Oklahoma. This type is said to be associated with pre-pottery cultures of the Archaic period.

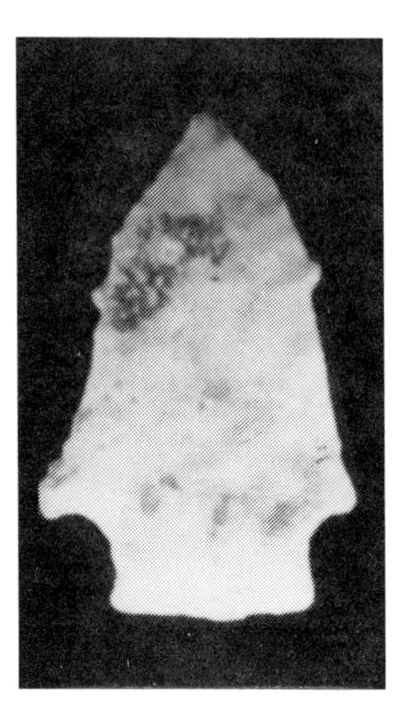

4-A - $50.00

Angostura

(Pictures 5-A thru 5-E)

The Angostura is medium to large point, with excurvate blade and incurvate base.

Angostura averages about 3 inches long and about 1-1/4 inches at the widest point.

The cross-section is biconvex. The blade shape is excurvate with some examples having slight shoulders at the terminal end of the shaft. Some examples may show beveling on one edge of each face. Most of this type have a contracted base, or pointed or rounded auriculate base. Though most will have incurvate basal edge, some may have straight or slightly excurvate base.

Eastern examples are usually shaped by broad random, or crude collateral flaking. Western types have fine oblique transverse flaking. Random is the much more common type of these two. First step of retouch was likely percussion, with a fine retouch as the last step.

(continued on next page)

The type was first recognized in 1949 by Hughes and called a Long Point from the Long Site in Angostura Resevoir, South Dakota. It was later given its present name. Limestone County examples have been estimated between 6,000 B.C. and 4,000 B.C. Angostura has been found in association with Lerma pointed base and a biface knife at the second Mammoth find in Santa Isabel Iztopan, Mexico. This example closely resembled Tennessee valley points, suggesting a late paleo or very early Archaic age.

5-A - $500.00

5-B - $75.00 **5-C - $100.00**

5-D - $35.00 **5-E - $35.00**

Arredondo

(Pictures 6-A thru 6-C)

A medium to large size point. Has an incurvate base with tapering shoulders.

Some examples resemble Savannah River points and is considered to be a middle or late Archaic point.

6-A - $200.00

6-B - $100.00

6-C - $75.00

Appalachian

(Pictures 7-A thru 7-D)

Medium to large, stemmed point, with a concave base.

This type ranges from about 2-1/2 to 4-1/2 inches long. The average width around 1-1/2 inches. They are often rather thick around 5/8 inches, because of the characteristics of working quartzite.

The shoulders are generally tapered and narrow. The cross-section is usually flattened. The blade is most often excurvate but is sometimes straight with acute distal end. The broad stem can be straight, slightly contracted or expanded. The basal edge is incurvate and often thinned. The stem is often ground around the edges.

The point is shaped by well executed percussion flaking, then finely retouched on edges.

This type was named by Kneberg in 1952. It was named for the southern Appalachian region were it is widely found. It seems the point is associated with Middle to late Archaic and early Woodland periods.

7-A - $35.00
7-B - $45.00

7-C - $25.00 **7-D - $25.00**

Autauga

(Pictures 8-A thru 8-E)

Autauga is a small, serrated, corner notched point with straight base and blade.

This type ranges from under 3/4 inch to about 1-1/2 inch length and around 1/2 inch to 1 inch wide.

Examples exist with either rhomboid or biconvex cross-section. The shoulders are horizontal or taper up toward the distal. Most examples are serrated and about half are steeply beveled. The blade edges are usually straight. The basal edge is straight and the notches are shallow. Both the corner notches and basal edge are lightly ground.

Shaping was performed by narrow, shallow, random flaking with deep flakes used to notch thus forming an expanded stem. The notches and base were finished by light grinding.

The point type was named by Cambron for Atauga County, Alabama. All these points were made of vein quartz. They are found with Dalton, Big Sandy, Kirk Corner Notched and Crawford Creek points.

(continued on next page)

Its similarity to these points seems to indicate an Early Archaic or possibly Transitional Paleo date.

8-B - $45.00 **8-C - $25.00**
8-A - $100.00
8-D - $25.00 **8-E - $25.00**

Bakers Creek

(Pictures 9-A thru 9-E)

Bakers Creek is a medium size point with an expanded stem.

This type ranges from less than 1-1/2 inches to over 3 inches. Though occasionally longer examples occur, the average will be around 2 inches or so. Average width is around 1 inch of less.

A cross-section of the point will be biconvex. The blade is generally straight but may be slightly exacurvate with a very acute distal end. The shoulders may be straight or slightly tapered and are usually shallow. The base is formed by notching about one third of the way from the base to the distal end. The base tapers from the shoulder to the basal edge, expanded at bottom out to about the width of the blade. Basal edge is thinned and often ground.

(continued on next page)

The point is shaped by random, broad, percussion flakes. Short deep flakes were removed to form the hafting area, then the edges all around were retouched for the final shape.

This point type was named by Cambron, after a site at the mouth of Baker's Creek in Morgan County, Alabama. Bakers Creek has been found in strata with Copena and Copena triangular. It was for this reason the point was called the Stemmed Copena in 1958. Evidence indicates an early to mid Woodland association. Estimated age is around 1,500 B.C. to early centuries A.D. It was changed to Bakers Creek around 1962.

9-A - $450.00 9-B - $250.00 9-C - $50.00
9-D - 25.00 9-E - $15.00

Beacon Island

(Pictures 10-A thru 10-E)

A medium size point with an expanded round stem. The blade edges are straight with prominent to weak shoulder barbs.

This type ranges from 2 inches to 3 inches on average. Width from 3/4 inch to 1-1/4 inches is about average.

The point has a biconvex cross-section. The shoulder may be barbed or taper downward. Blade edges may be finely serrated and are straight with acute distal end. The stem round and expanded having diagonal notches at the shoulder.

The shaping of the faces were by shallow random flaking followed by careful pressure flake retouch to shape and thin the base.

The point was named for Beacon Island by Allen and Hulse. It is found with Cotaco Creek and Flint Creek points, suggesting a late Archaic to early Woodland occurrence.

10-A - $45.00 **10-B - $75.00** **10-C - $35.00**
10-D - $25.00 **10-E - $15.00**

Beaver Lake

(Pictures 11-A thru 11-E - Photos Reduced)

Beaver Lake is an auriculate point with recurvate blade edges. This medium sized point often exhibits extremely fine workmanship. This type ranges from less than 1-1/2 inches to over 4 inches, around 2-1/2 inches being average. It should be noted pieces much over 3-1/2 inches would be very rare. Width is usually around 1 inch to 1-1/4 inches.

The cross-section may be medium ridged but is usually biconvex. The blade is recurvate, with the haft constricting just above the auricles. The auricles are expanded and rounded or slightly pointed. This gives the base a fish-tail appearance. The entire haft and incurved base are usually well ground. Shaping of the blade was accomplished by shallow random flaking, sometimes resulting in a median ridge. Secondary flaking is usually long, evenly spaced, and struck from alternating faces. This retouch was performed by using pressure flaking.

This type was named for the Beaver Lake area in Limestone County, Alabama by Cambron and Hulse. Examples were recovered from the bottom strata at several control digs in north Alabama. This indicates a 10,000+ history. Most examples are from Ft. Payne, Chert, and are well patinated.

11-A - $1,000.00 (3-1/2")
11-B-$500.00- 11-C-$350.00 11-D-$200.00 11-E-$200.00

Benjamin

(Pictures 12-A and 12-B)

Benjamin is large to medium size point with excurvate blade edge and excurvate basal edge.

The type ranges from 2-1/2 inches to over 3-1/2 inches in length and around 1 inch in width.

Cross section is biconvex. The blade is excurvate with a poorly defined hafting area. The widest point may either be about mid-blade or at the base. Basal edge tends to be excurvate and thinned but may be straight. The distal end is acute.

Blade faces are shaped by deep, broad random flakes with fine retouch used lightly around the blade edges. The basal edge was thinned by a series of short random flakes.

This type was named by Cambron, after the Benjamin site in Lawrence County, Alabama. Here it was found in a Woodland culture association. It has been found in conjunction with Copena and Madison points. They are generally made of local materials.

12-A - $45.00 **12-B - $20.00**

Benton Broad Stemmed

(Pictures 13-A thru 13-E)

Benton Broad Stemmed is a medium to large point, with broad stem with steeply beveled edges.

Average length of type is around 2-1/2 inches and from 1 inch to 1-1/4 inches wide. The Broad Stemmed type appears much wider for its length than the usually longer Benton Stemmed.

The cross-section is either flattened or biconvex. The blade edges are usually excurvate but sometimes have parallel straight edges just above the shoulders for 1/3 or more of the length. Shoulders are straight to slightly tapered. The stem tends to be slightly expanded and steeply beveled all the way around, and slightly incurvate, straight or excurvate. This stem is very short and wide.

The blade was shaped by broad, shallow random flaking and finished with broad retouch. Base was formed by removing corners from basal edge. The steepness of the flaking around the stem is a strong indicator for this type.

The name was derived from Benton County, Tennessee by Kneberg in 1956. The type is found widely in the Tennessee River Valley and in Western Tennessee along the Mississippi River. Association with Big Sandy and Eva and examples recovered in controlled digs lead to a 4,000 B.C. to 2,000 B.C. Archaic period occurrence. Some believe Benton to be even earlier. Not as commonly encountered as Benton Stemmed.

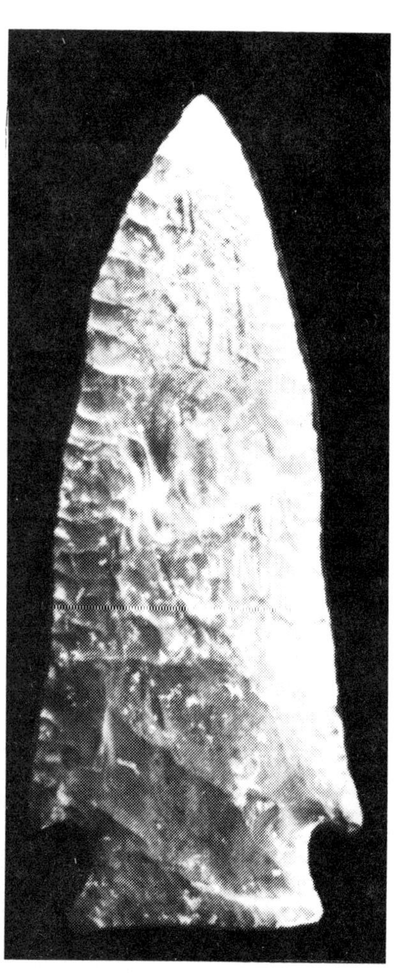

13-A - $500.00

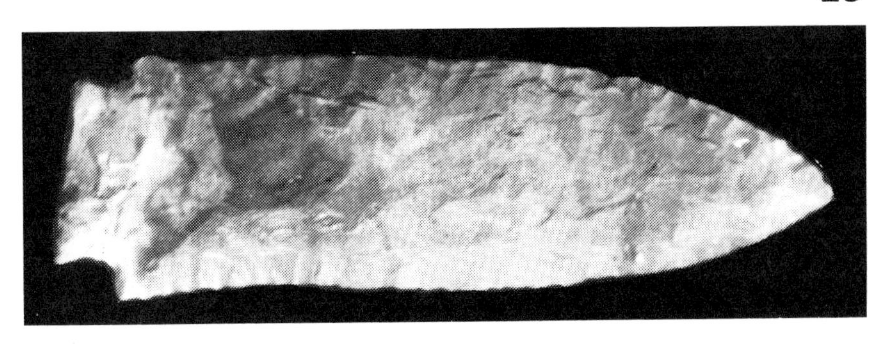

13-B - $175.00

13-C - $100.00 13-D - $50.00 13-E - $75.00

Benton Stemmed

(Pictures 14-A thru 14-D)

Benton is medium to large, stemmed with a steeply beveled base.

This type may range from under 2 inches to over 4 inches. The width averages about 1 inch to 1-1/4 inches.

(continued on next page)

Benton will have a biconvex cross-section. The blade may be recurvate to slightly excurvate with an acute distal end. Shoulders are horizontal or with a slight taper. The shoulders are narrow or rarely barbed. The stem is relatively broad but not as broad as in Benton Broad stemmed. It also tends to have a longer narrower appearance over all, than the Broad Stemmed type.

Random, broad and shallow flaking was used to shape the blade with retouch along the blade edges being broad and shallow. Short steep flakes formed the base and appear to have been pressure flaked. It may be of interest to note there have been a number of these points made by rechipping blades from older more heavily patinated blades.

Benton was named by Kneberg in 1956. It was named for sites in Benton County, Tennessee. Distribution is widely spread through the Tennessee River Valley and Western Tennessee along the Mississippi River. This type is more often encountered than the Broad Stemmed Benton.

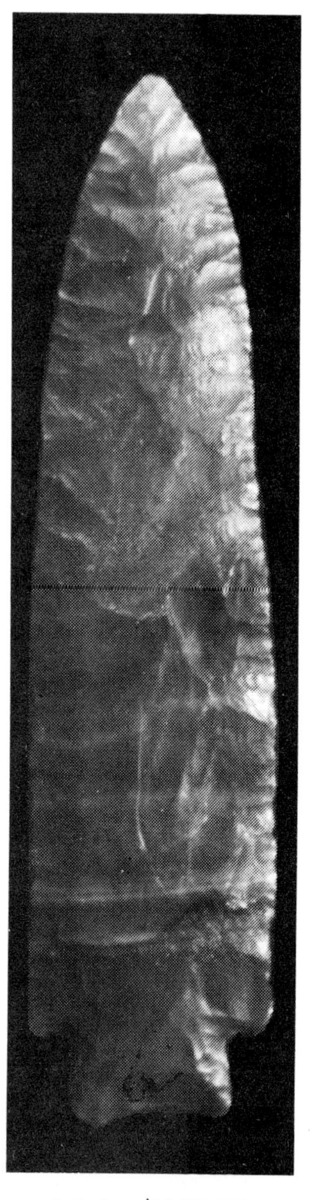

14-A - $500.00

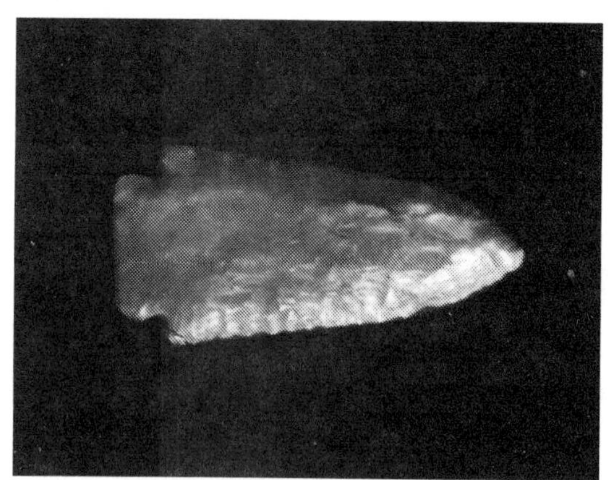

14-D - $35.00

14-B - $250.00
This point was stolen ...
contact author if seen!
Size Reduced - Actual Size 5"

14-C - $100.00

Big Sandy

(Pictures 15-A thru 15-N)

Big Sandy is a small to medium point with distinct side notches. Blade may be beveled and/or serrated. Basal edges and notches are often ground.

Big Sandy may be from under 1 inch to around 2-1/4 inches in length. The width usually averages around 3/4 inch at the shoulder. Later unground examples tend to be longer points with some examples in excess of 3 inches long.

(continued on next page)

The flaking on Big Sandy is quite varied but usually well worked. Random flaking was generally used for primary shaping of the blade and hafting area. Fine retouch is usually evident around the edges of the blade and base with the base often being ground in the notches and basal edge. Some points are finely serrated and or beveled on opposite alternate faces.

This type varies greatly from point to point. The cross-section may be biconvex, plano-convex, rhomboid or median ridged with biconvex being the most prevalent. Some are beveled on one face on each side and a smaller number are serrated. The distal end is usually acute but there are a good number found with a scraper tip.

The base area below the notches may be squared or auriculate, rarely with pointed parallel basal edge.

The type is named for the Big Sandy I phase of the Archaic period, named by Lewis and Kneberg in 1959. Big Sandy is found in association with Dalton, Wheeler and Quad points suggesting a beginning during the transititional Paleo period. Carbon-14 dating from the Stanfield-Worly Shelter indicate 10,000 B.C. date. The Big Sandy may have continued until about 1,000 B.C.

Variants are: Big Sandy Ariculate; Big Sandy Broad Base (often confused with Osceola); Big Sandy Contracted Base.

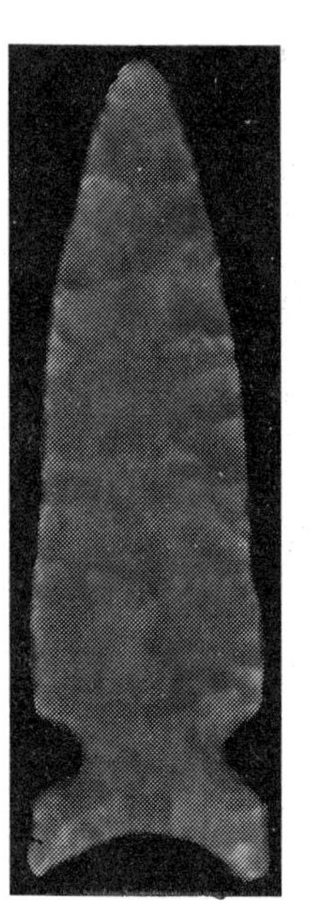

15-A - $150.00 Graham Cave Point

15-K - $30.00 15-L - $75.00 15-M - $30.00

15-B - $100.00

15-C - $125.00 15-D - $50.00 15-E - $50.00 15-F - $75.00
15-G - $75.00 15-H - $50.00 15-I - $45.00 15-J - $50.00
15-N - $200.00

Big Slough

(Pictures 16-A thru 16-E)

Big Slough is a medium to large point with a broad expanded stem.

This type ranges from less than 2 inches to over 3-1/4 inches in length, and from 1 inch to 1-1/2 inches wide at the shoulders.

Cross-section is biconvex. The blade may be either excurvate or recurvate, only rarely excurvate-recurvate. The shoulders are inversely tapered and barbed, sometimes being expanded. Distal end may be apiculate but is usually acute. The stem is broad, long and expands toward the basal edge. The stem edge may be slightly incurvate, the basal edge is thinned and usually ground.

Most of this type was shaped by broad random flaking but a significant number show collateral. Sometimes one face will be collateral and the opposite face is random on the same point. Blade edges generally show fairly broad retouching, with fine chipping over it on the edges. The basal area was thinned with broad shallow flakes then finely retouched and ground.

Named for the Big Slough Area in Limestone County, it was first recognized by Cambron in 1960. Probable age of about 5,000 B.C. to 2,000 B.C. indicate an early to middle Archaic association.

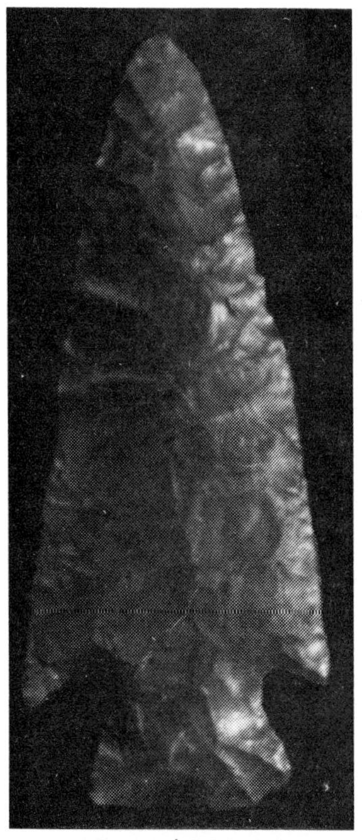

16-A - $200.00

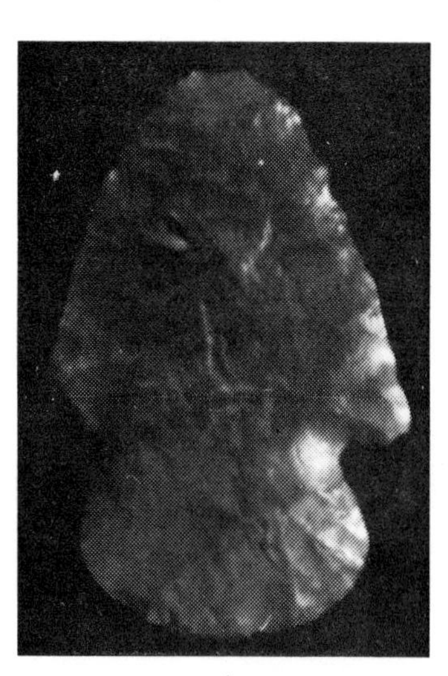

16-F - $35.00

16-B - $200.00 16-C - $250.00

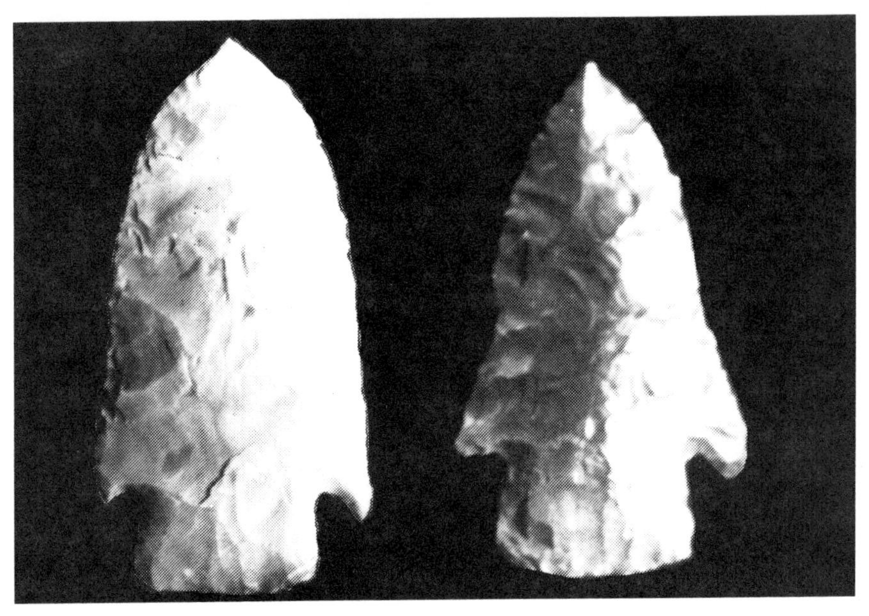

16-D - $100.00 16-E - $100.00

Boggy Branch

(Pictures 17-A thru 17-C)

Originally thought to exist only in South Alabama and Georgia - the type has been found as far north as the Tennessee River Valley.

Two types exist - type one has an elongated bulbous stem. Type two has a short ovate stem. The point is usually serrated and the base is ground. Thought to be associated with the Kirk or Big Slough cultures.

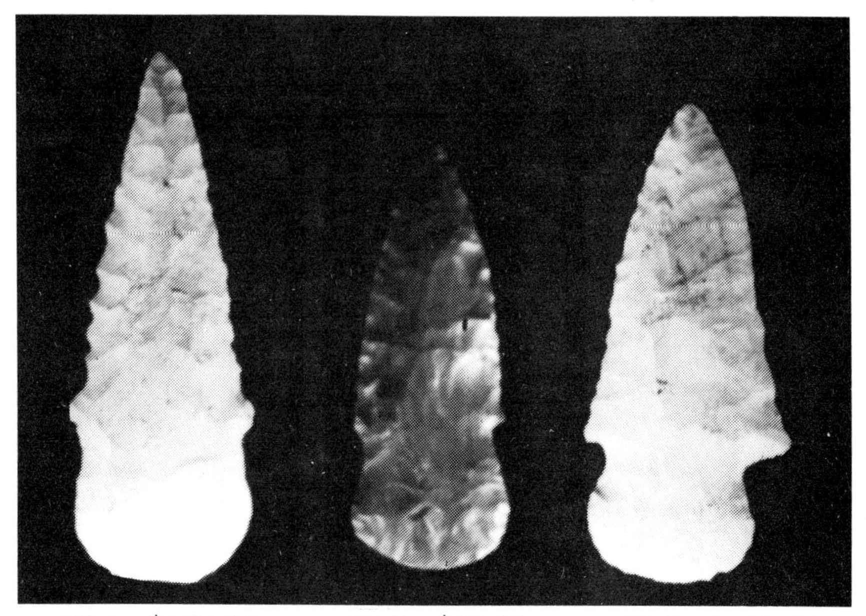

17-A - $150.00 **17-B - $75.00** **17-C - $100.00**

Bolen

(Pictures 18-A thru 18-G)

A cluster of side and corner notched points found in Florida, South Georgia and South Alabama. Variants are Bolen Side Notched, Bolen Expanded or "E" Notched and Bolen Corner Notched (which may be a Plevna Variant) and Bolen Recurvate has a recurvate basal edge instead of the excurvate or straight base of the other varieties. The Bolen is an early Archaic point type.

(points on following page)

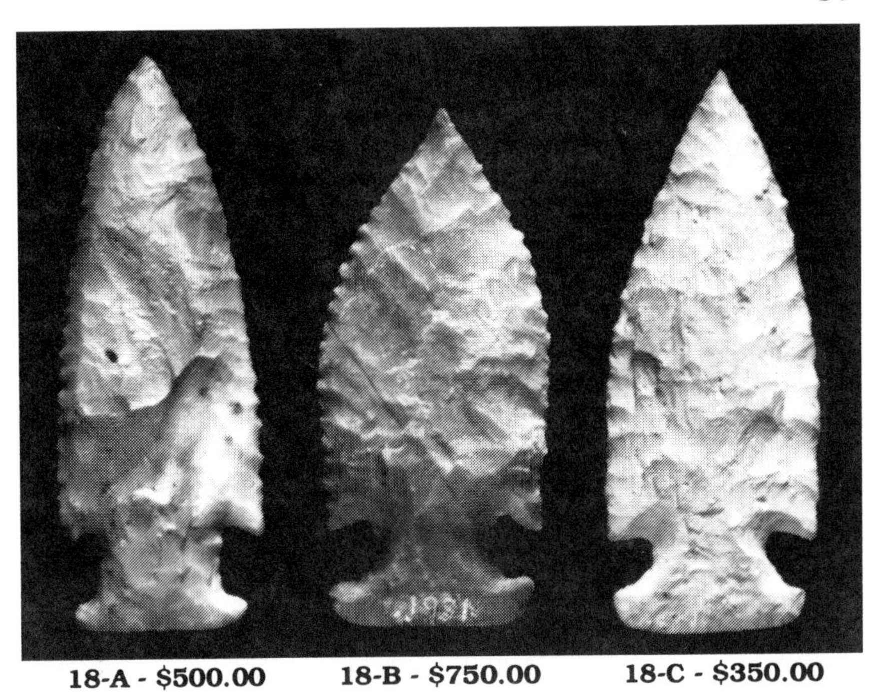

18-A - $500.00 18-B - $750.00 18-C - $350.00

18-D- $300.00 18-E - $400.00 18-F - $300.00 18-G - $100.00

Bradly Spike

(Pictures 19-A thru 19-F)

A medium to small spike shaped, stemmed point. Average length ranges from 1-1/2 inches to 2-1/2 inches in length and a width of around 1/2 inch.

Usually the cross-section will be median ridged but in rare instances it may be plano-convex or biconvex. The shoulders taper and may be asymetrical. Blade may be straight to slightly convex. The hafting is usually a straight stem but may taper. Stem sides are straight with an excurvate or straight basal edge.

Random flaking was used to form the blade. This flaking was steep and blade was often almost as thick as it was wide. At times the flaking was struck from alternate faces giving an irregular outline to the blade. Fine retouch was not generally used resulting in most points having a somewhat crude look.

`The point was named for Bradley County, Tennessee, where it was first recognized by Kneberg in 1956. This type appears associated with Woodland culture with a date beginning around 2,000 B.C. to sometime A.D.

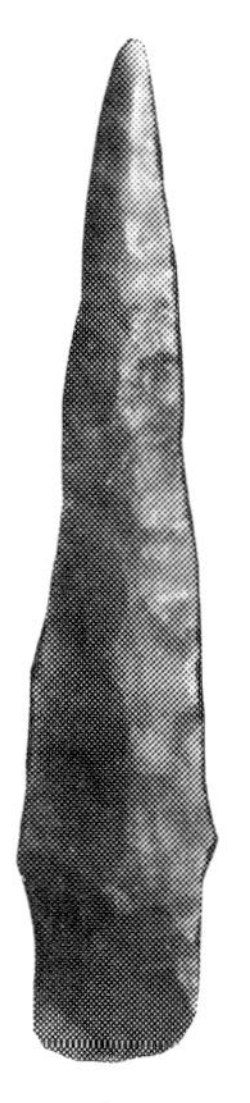

19-A
$75.00

19-C-$20.00 19-D-$20.00 19-E-$15.00 19-F $35.00

19-B
$45.00

Buzzard Roost

(Pictures 20-A thru 20-E)

This is a medium to large, bifurcated-stemmed point.

Length averages from 2-1/2 inches to over 4 inches with an average width between 1 inch and 1-1/2 inches. Example are known over 4 inches but are not commonly seen.

The cross-section is biconvex. Shoulders are often inversely tapered with a slight barb expansion. The blade is recurvate. The distal end is usually acute. The stem may be straight or slightly expanded with a bifurcated or sometimes auriculate appearance. The stem edges are usually steeply beveled.

The shaping of the blade employed random, broad and thin flaking, or rarely collateral flaking. The edges of the blade were finished with long shallow pressure flaking. Short deep pressure flaking was used to form the base with the typical steep bevel.

The type derives its name from Buzzard Roost Creek in Colbert County, Alabama and was first recognized by Cambron. The point appears to be related closely to the Benton Stemmed point. An appearance around early to middle Archaic seems likely.

20-A - $500.00
This point was stolen ... notify author if seen!

20-B - $50.00

20-C - $125.00 20-D - $35.00 20-E - $100.00

Cahokia Point (Tri Notch)

(Pictures 21-A thru 21-D)

Early to late Mississippian period. Similar to Madison Point, except for the notching. Notching is characteristic double notch, triple notch or multiple notch and serrated. Distribution: Illinois, Iowa, Wisconsin, Missouri, North Arkansas and Oklahoma.

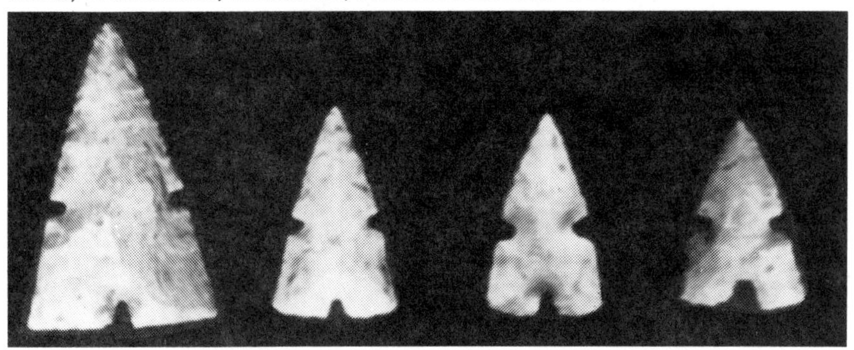

21-A - $150.00 21-B - $75.00 21-C - $75.00 21-D - $45.00

Camp Creek

(Pictures 22-A thru 22-F)

Camp Creek is small to medium sized point, with triangular shape and incurvate base. Length may range from 1 inch to slightly over 1-1/2 inches. The width averages around 5/8 inch at the widest point.

Cross-section is generally biconvex. The blade is most often straight but may be slightly excurvate or incurvate. The base is usually thinned with some fine retouch. The distal end is usually acute.

The blade was shaped by fairly well executed random flaking. Even though fine retouch was employed the point often has a crude look because of the often poor quality of flint or quartzite used.

The point derives its name from the Camp Creek site on the Nolichucky River in Greene County, Tennessee. It was first recognized by Kneberg in 1956. It is usually found with early to middle Woodland culture associations. Kneberg suggested an age of about 1,000 B.C. to 500 A.D.

(points on next page)

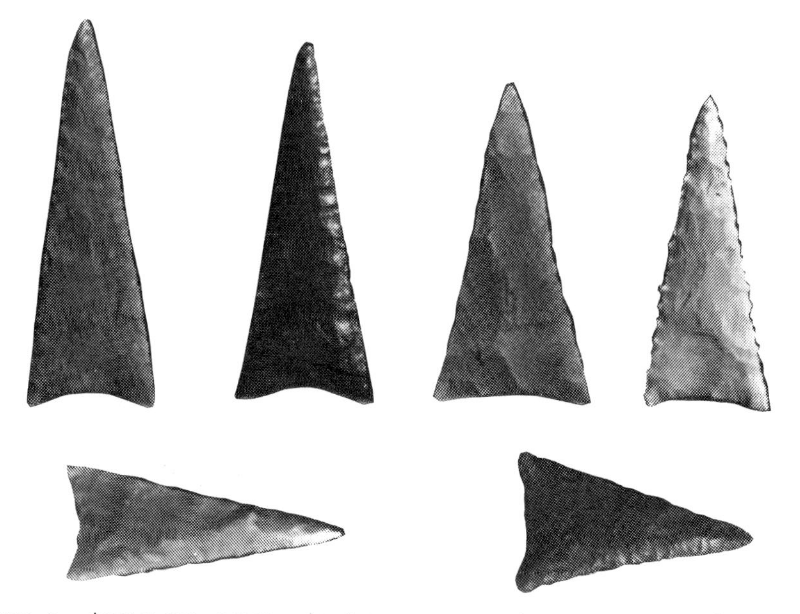

22-A - $150.00 22-B - $75.00 22-C - $30.00 22-D - $100.00
22-E - $25.00 22-F - $35.00

Candy Creek

(Pictures 23-A thru 23-E)

Candy Creek is an auriculate point of medium size. It has recurvate edges and an incurvate base. Average length is around 2 inches and average width at widest point is around 3/4 inch to 1 inch.

The blade is recurvate with biconvex cross-section. The distal end is acute. Hafting area is auriculate with expanded rounded ears and an incurvate basal edge. The base is sometimes thinned and hafting edges may be lightly ground.

The blade is shaped by shallow random flaking with fine retouch used to finish the blade and haft. There is thinning on the basal edge that can sometimes look like the flute on Paleo points. This is not considered related to Paleo fluting but rather as an accidental result of the basal thinning process, resulting in broad shallow scars.

The name comes from Candy Creek site in Bradley County, Tennessee, and was first recognized by Kneberg in 1956.

(continued on next page)

These points appear to be very closely associated with Copena and Copena Triangular types, sharing both a similarity of workmanship and materials used. A date of about 1,000 B.C. to 500 A.D. during the Woodland period.

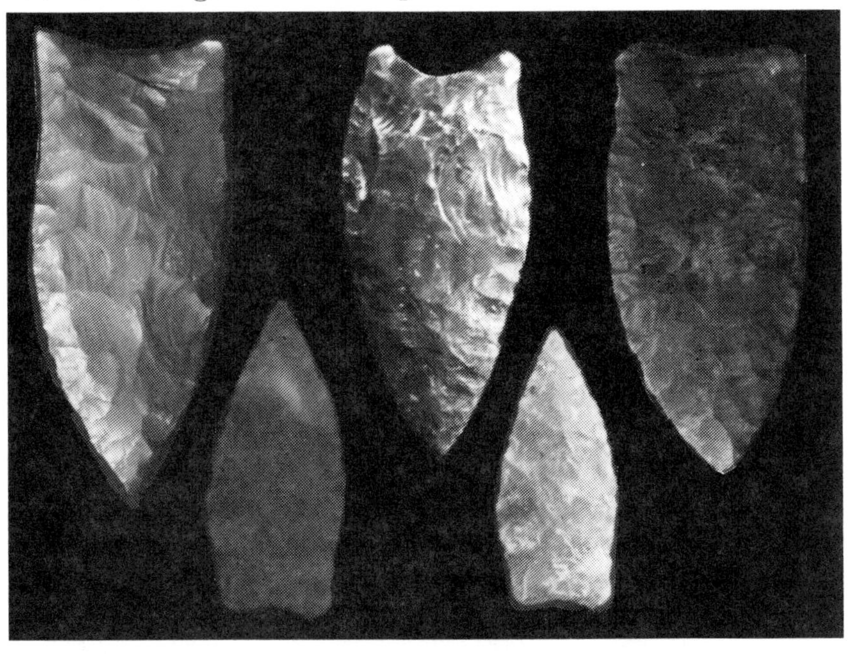

23-A - $50.00 **23-B - $35.00** **23-C - $20.00**
23-D - $15.00 **23-E - $20.00**

Cave Springs

(Pictures 24-A thru 24-E)

Cave Springs are medium to small points with a bifurcated stem. Length ranges from about 1 inch to 1-1/2 inches with a width at shoulders of just under 1 inch to 1/2 inch.

The cross-section most often is biconvex but may be plano-convex or medium ridged. Shoulders are generally tapered but may be barbed or horizontal. Blade edges are usually straight but on rare occasion can be excurvate. The acute distal end is most common but examples are known with a broad point. The stem is shallowly bifurcated and expanded usually wider than its length.

(continued on next page)

The blade and haft were shaped by broad, shallow, random flaking. The blade was usually finished with collateral flaking. Fine retouch is widest on most blade edges. Long shallow flakes were used to form the basal bifurcation. One large flake was stuck from each side of each face to form a shallow notch that separates blade from stem.

The type is named for the Cave Springs site in Morgan County, Alabama. First recognized by Cambron, examples were recovered from the excavation of Cave Springs site. Blades of this type resemble Big Sandy. A transitional Paleo to early Archaic is suggested.

24-A	24-B	24-C	24-D	24-E
$35.00	$35.00	$35.00	$10.00	$10.00

Citrus

(Picture 25)

Similar to the Culbreth except that its basal edge is excurvate while the Culbreth is straight. The Citrus descended from the Culbreth. A gulf formational type.

(point on next page)

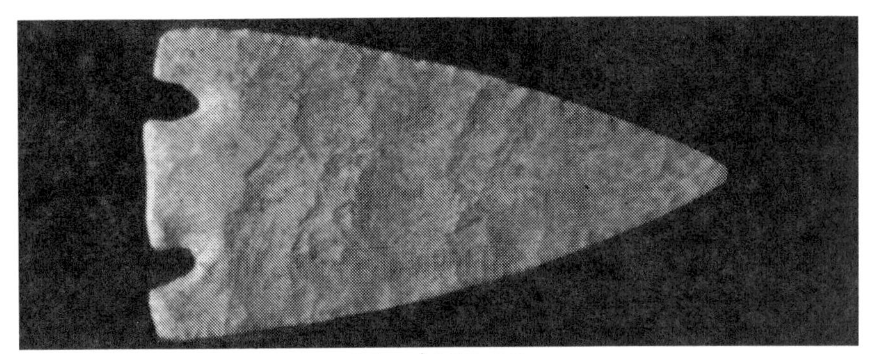

25 - $350.00

Clay

(Pictures 26-A thru 26-B)

One of the only two notched forms appearing in the coastal plain region during the late Archaic period. Blade edges are recurvate to incurvate. Barbs are squared off, sometimes rounded. Notches are usually long and narrow. The cross-section is thin.

26-A - $500.00 **26-B - $450.00**

Clifton Blade

(Pictures 27-A thru 27-E)

A medium to large pointed oval blade. It is a part of the Jacks Reef Culture. Blade length runs from 3-1/2 inches to over 9 inches. Widest part of the blade is usually 1/3 of the way from the base. The blade is always thin and well made.

It resembles the Adena blade but has a rounder basal edge than the Adena blade. Occasional rare examples have a small flat basal edge. A few examples have been found with a pentagonal outline.

It is found on Woodland sites and is usually made of an almost white heat-treated chert and usually has streaks of pink mingled in. Nearly all of them appear to have been made of material from the same quarry. This type has a limited range and is found mostly along the Tennessee River Valley from western Alabama to northern Tennessee.

Group 27-D is an assemblage of artifacts found in direct association with each other and is the first time Clifton Blades have been found in cultural relationship with other artifacts. Notice the arrow points are Jacks Reef Pentagonals.

27-D - Total Price $500.00
Blade at Top - 5"

27-A	27-B	27-C
7-1/2" , $2,000.00	5-1/2", $1,000.00	7-1/4"$1,800.00

27-E

A cache of Clifton Blades - Value $4,500.00

Longest - 5" (A cache of these blades are very rare)

Clovis

(Pictures 28-A thru 28-L)

A fluted point - ariculate in form with an incurvate rarely straight base. Average size is 2 to 3 inches in length and one inch in width.

The blade is usually excurvate, seldom recurvate. The distal end most often acute sometimes broad. The auriculate base may be pointed or rounded. The sides of the hafting area are usually parallel but may be contracted.

Most examples are fluted one third the length of the blade. Some may be fluted the entire length of the blade. Flutes may be single or multiple. Most examples are ground on the base. Flaking is random but on better examples may be collateral. All examples are finished by retouching on the edges. Flutes were stuck by direct percussion after preparation of a "striking platform or nipple" on the base.

The point is named for Clovis, New Mexico where examples were found in association with mammoth bones. Examples are usually found in association with uniface blade tools.

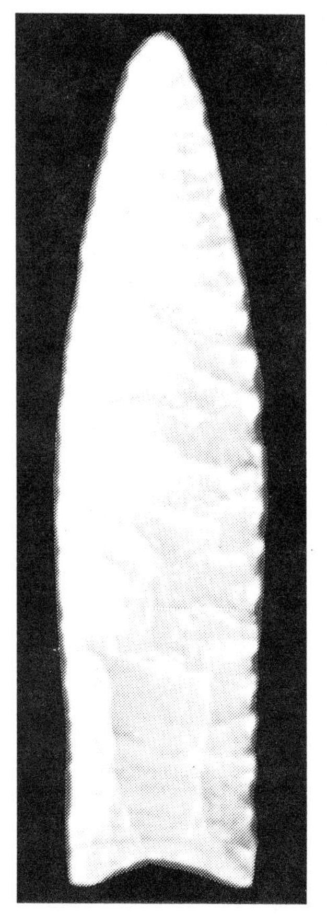

28-A - $1,000.00
Bone Saw

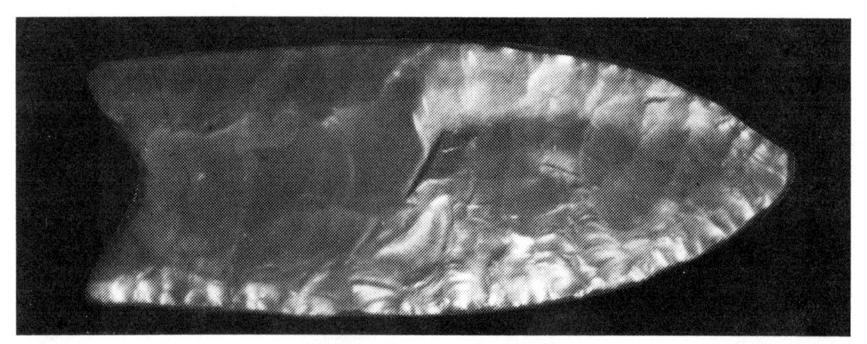

28-B - $1,200.00

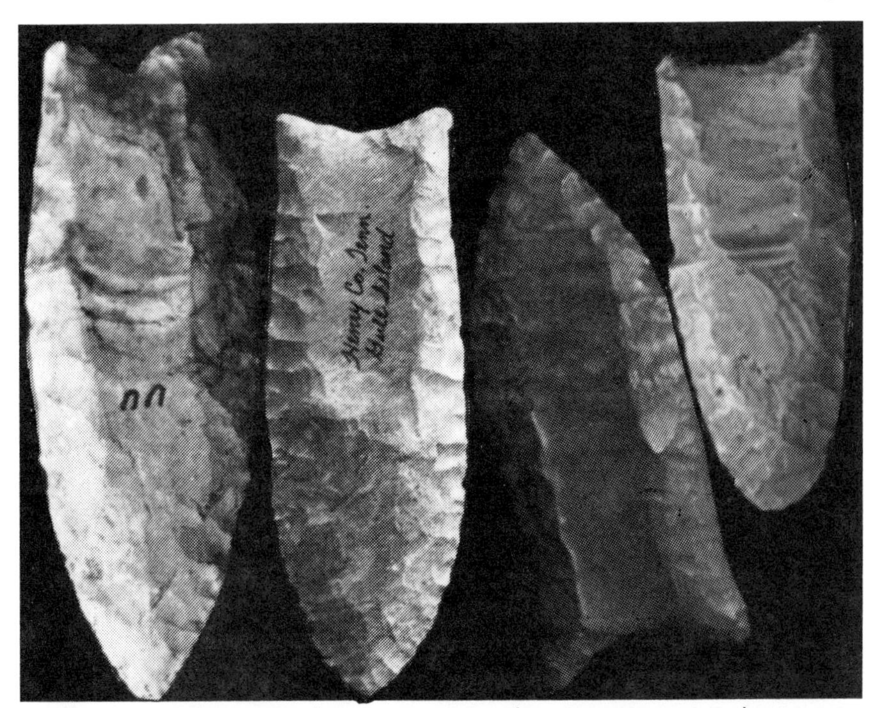

28-C-$500.00 28-D-$750.00 28-E-$1,000.00 28-F-$250.00

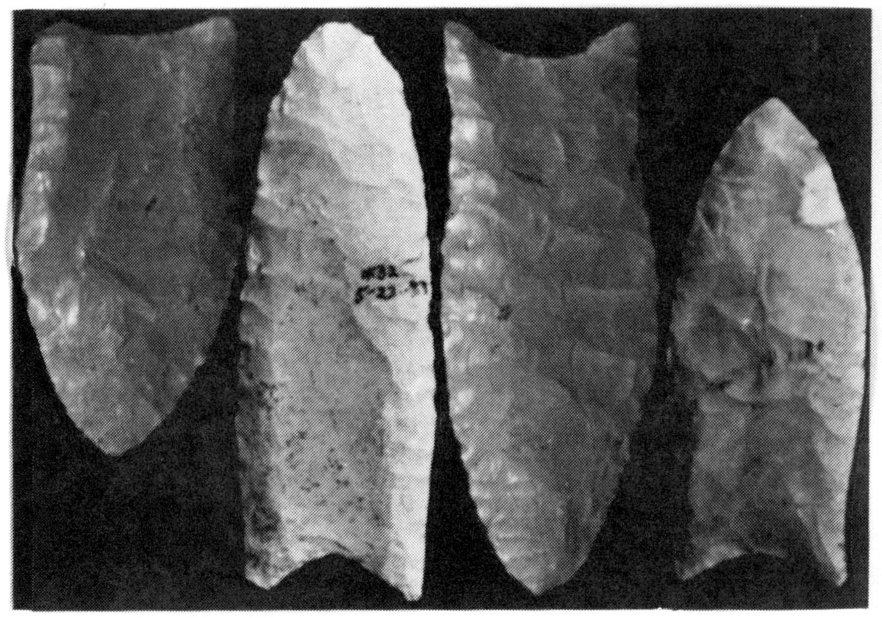

28-G-$200.00 28-H-$400.00 28-I-$250.00 28-J-$200.00

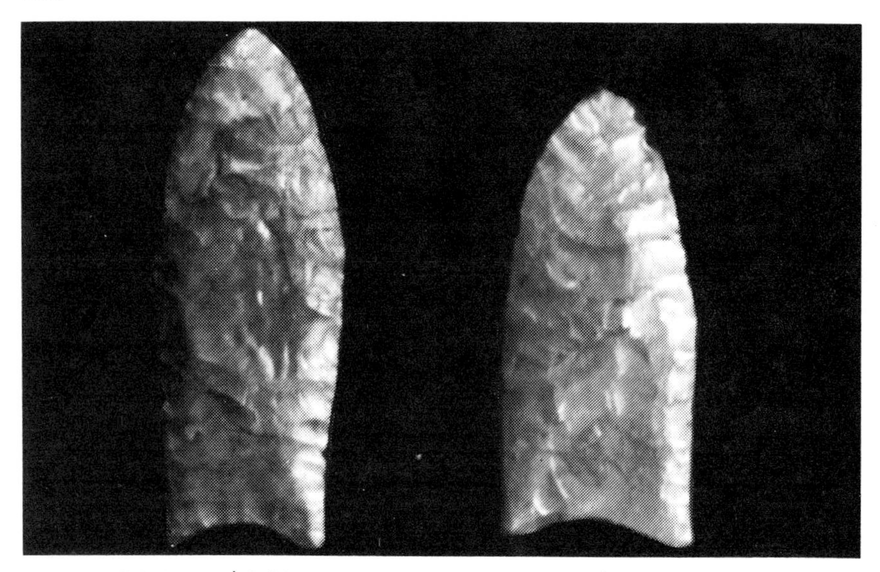

28-K - $250.00 **28-L - $175.00**

Clovis, Unfluted

(Pictures 29-A thru 29-C)

With exception to not being fluted, the information is the same as for the fluted Clovis.

Basal edge is thinned by retouching the same as the blade edges. Some of the examples often identified as unfluted Clovis may actually be unsharpened Dalton points.

(points on following page)

29-A - $250.00 **29-B - $75.00** **29-C - $150.00**

Cobbs

(Picture 30-A thru 30-E)

This blade is the knife form or preform for the Lost Lake and Plevna types. Straight base types are associated with Lost Lake - round base ones with the Plevna. Several examples show either one notch or the preparation of the base for notching by thinning the corners of the blade. Some examples show definite use as knives.

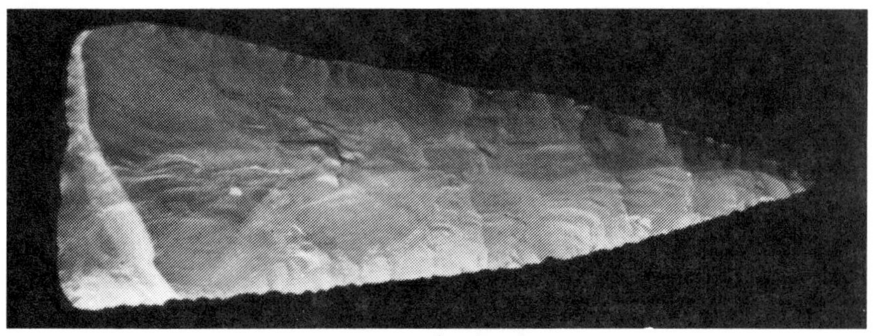

30-B - $750.00
Actual Size 4-1/2" - Pickwick Tri-Color Jasper

30-A - $300.00 **30-C - $75.00** **30-D - $250.00**
30-E - $100.00

Coldwater

(Picture 31-A thru 31-D)

This is a stepped base point. The base has the appearance of being stair-stepped or of having one small stem on top of a wider stem. It may have some cult association as the same pattern is found on the rim of some pottery. A late Woodland association is assumed. The type is named after the historic trading post of Coldwater on the Tennessee River in Colbert County, Alabama. Several of this type have been found in the surrounding area. The type is not common and is found over much of the southeast.

31-A - $50.00 31-B - $45.00 31-C - $35.00 31-D - $75.00

Conerly

(Picture 32-A)

Conerly is a medium to large sized point with incurvate base that is contracted. Conerly ranges from about 2-1/4 inches to over 4 inches in length and between 1-1/4 inches and 3/4 inch in width with 1 inch being average.

Cross-section may be flattened but is usually biconvex. The shoulders may be defined very weakly if at all. The blade edges are usually shallowly serrated and either straight or excurvate, with an acute distal end. The hafting area is contracted with incurvate basal edge.

Though flaking is usually random and fairly well performed, it may rarely be transverse oblique. Most examples are roughly beveled for about one fourth the width of the face, this retouch is responsible for the bevel also leaves some fine serrations on the edges. The base was thinned after the concavity was flaked in.

The type was named for the Conerly site near Sardis in Burke County, Georgia. Physical appearance of this type is similar to Savannah River points as well as Guilford points. There seems to be a likely association with the Archaic period. The type was identified by Lively.

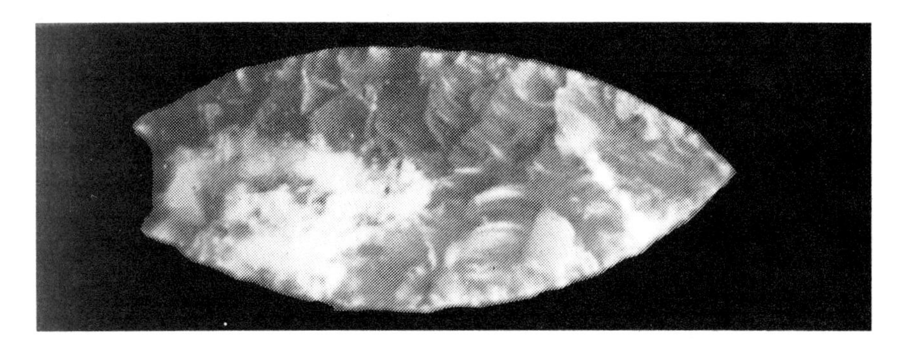

32-A - $125.00

Copena

(Pictures 33-A thru 33-F)

Copena is a medium to large point with a trianguloid shape and recurvate blade. Examples range from 2 inches to over 4 inches. Points of over 4 inches are uncommon but may rarely be as much as 6 inches. Width will average around 7/8 inch to 1-1/4 inches in longer points.

Copena has a biconvex cross-section. The constriction at the haft area results in a recurvate blade. The distal end is usually acute but may be excurvate. The hafting area is defined by a constriction just above the basal edge. This haft flares out at the basal end and basal edge is excurvate. The hafting edges are usually lightly ground and basal edge usually thinned.

Shaping the blade employed broad, shallow, random or collateral flaking. Some exhibit a weak median ridge. Edges are finely retouched by pressure flaking. Material varies widely, as local flint was used.

This type was named for the Copena burial mound culture in north Alabama. "Copena" is a contraction of the two materials commonly found in burials of this culture, Copper and Galena. Copena is often referred to as Southern Hopewell. The point is usually found in burial mounds of a Woodland association, but is sometimes associated with late Archaic sites in the Tennessee area. Suggested age around 500 B.C. to A.D.

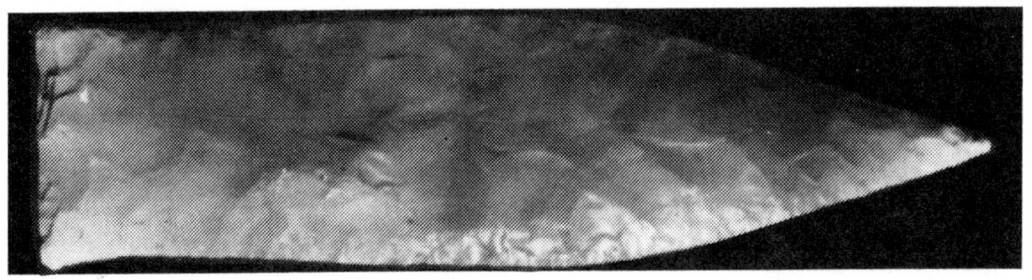

33-C - $500.00

33-B - $125.00

33-A- $150.00 33-D - $350.00 33-E - $75.00 33-F $25.00

Copena Triangular

(Pictures 34-A thru 34-E)

This is a medium to large trianguloid point with a parallel sided haft. Examples measure from 1-3/4 inches to about 3 inches in length and about 3/4 inch to 1-1/4 inches in width.

Cross-section is almost always biconvex or rarely plane-convex. The sides of the haft are nearly parallel with excurvate blade above the haft, but may flare out slightly at the base.

(continued on next page)

The point usually has an acute point. The basal edge may be straight or slightly incurvate. All bases will be thinned with the incurvate ones usually ground. The straight base type may be ground on sides of hafting area.

Shaping of the blade and haft was started with random flaking, followed by secondary short broad flakes, finished by fine retouch pressure flaking. Basal edge was thinned by short broad and shallow flakes but in unusual examples flakes may approach the length of the haft area. Points are made generally from local sources of material.

The type is named for the Copena Burial Culture in north Alabama, first recognized at Cambron site 53 in Morgan County. Origin appears to be in Archaic lasting into the Woodland period.

34-A - $500.00 34-B - $50.00 34-C - $35.00 34-D - $35.00 34-E - $25.00

Cotaco Creek

(Pictures 35-A thru 35-G)

This is a medium to large point with round shoulders and straight stem. Cotaco Creek may be from 1-1/4 inches to over 5 inches in length.

Cross-section is usually flattened but on occasion may be rhomboid. The shoulders are usually rounded and straight but may taper slightly. The blade is more commonly straight but may be excurvate. The blade edges may be beveled on the right side of one or both faces, they may also exhibit fine serrations. The distal end is usually apiculate or obtuse, only rarely may be acute. The stem usually has a square shape with straight stem sides and basal edge. Excurvate basal edge or slightly expanded stem occurs but only rarely.

The blade and haft were shaped by well controlled, shallow, random flaking. Edges were refined by the removal of fine, short, well spaced flakes forming fine serrations. When one face was beveled longer, narrow and even flaking was used. Ft. Payne chert, Bangor nodular and other local materials were commonly used.

The type was named for Cotaco Creek in Morgan County, Alabama where a number of points have been noted. Examples excavated at numerous sites indicate a beginning in the Archaic period with a continuance well into the Woodland period. Note that a variant has been recognized called Cotaco Creek Variant. This point is usually smaller, cruder, often more narrow, and is more likely to be acute at the distal end and have an excurvate blade edge.

35-E - $40.00 **35-F - $35.00** **35-G - $15.00**

35-A - $250.00

35-B - $125.00 35-C - $75.00 35-D - $75.00

Crawford Creek

(Pictures 36-A thru 36-C)

This is a medium-sized stemmed point which has a straight blade with fine serrations. Averaging from 1 inch to 1-3/4 inches in length and a width range from 3/4 inch to 1-1/2 inches.

This type has a biconvex cross-section. The shoulders may be horizontal or tapered with straight blade edges. Nearly all examples are finely serrated. Distal ends are usually acute. The stem is usually straight but may be expanded. Stem sides may be either straight or incurvate. Strong thinning is almost always applied to the basal edge which is usually straight but may be slightly incurvate. Some evidence indicates the expanded stemmed types are older than the straight stem types.

Blade and stem are formed by irregular random flaking, leaving some large blade scars showing on the faces. Deep and broad flakes were removed to shape the haft area, then broad and shallow flakes were removed to thin the basal edge. Retouch along the blade was done with steep, fine, pressure flakes. Serrations were formed by removal of small flakes from alternating faces.

This type was named for Crawford Creek in Morgan County, Alabama, and originally was called Provisional Type 3, before naming by Cambron and Hulse. A likely date of 5,000 B.C. or early Archaic into later Archaic.

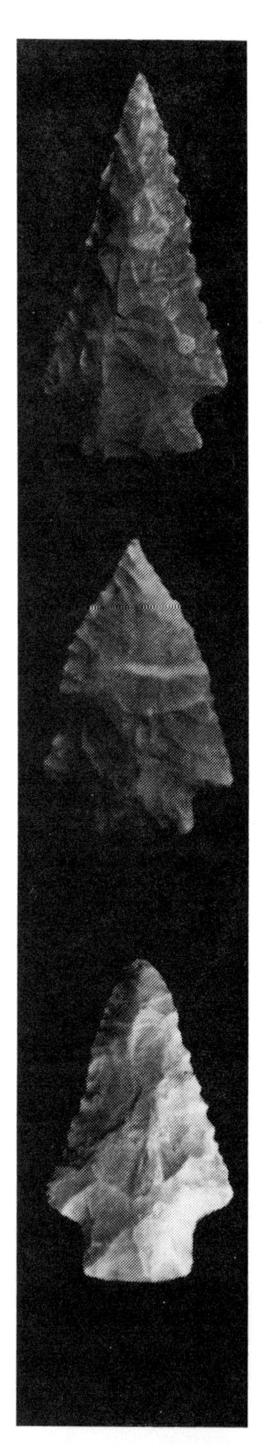

36-A
$35.00

36-B
$50.00

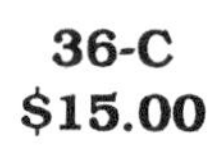

36-C
$15.00

Culbreth

(Pictures 37-A thru 37-B)

Similar in form to the Eva, but is not as old and probably is not related. Basal edges sometimes lightly ground. Found in Florida and surrounding areas.

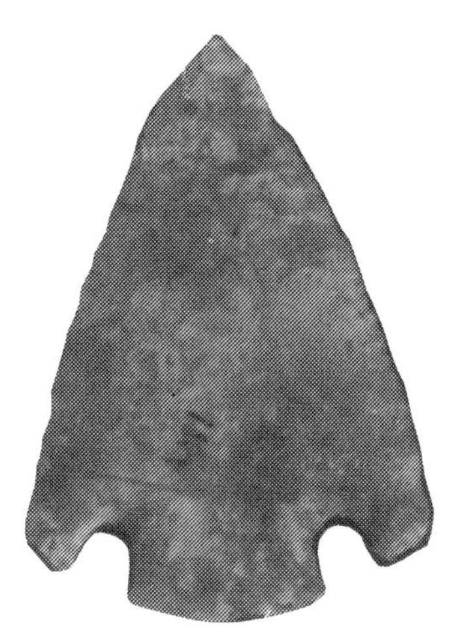

37-A - $100.00

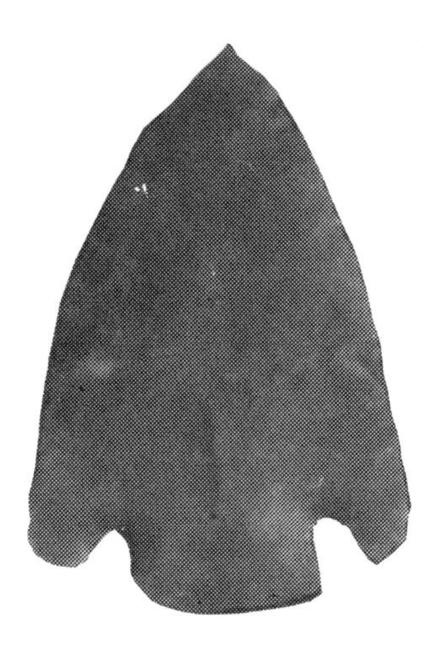

37-B - $125.00

Cumberland

(Pictures 38-1-A, 38-A thru 38-H)

A medium to large fluted point. Shape is recurvate with an expanded rounded auriculate hafting area. Size is 1-1/2 inches to 6-1/2 inches, with the average being 3 inches.

Examples of more than 4 inches are very rare. (One notes that in the flea market trade examples of 4 to 5 inches are plentiful.)

Fluting usually is not as wide as that on the Clovis but extends further. Most examples are fluted at least 1/2 the length of the blade, some even extending to the distal end.

Cross-section before fluting was medium ridged. There is no break between hafting area and blade as in the Dalton group. The hafting area is almost always ground. Fluting was accomplished by the direct percussion of a fluting nipple on the basal edge. After fluting the basal edge and hafting were finished by retouching.

This type was named after the Cumberland River Valley by Kneberg in 1956.

38-H - $150.00

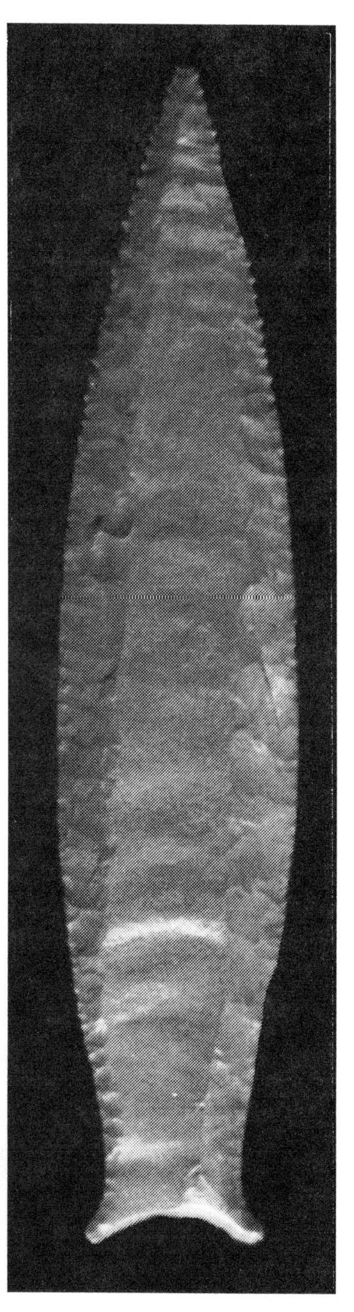

38-1-A
$19,000.00

38-B - $1,000.00
38-C - $1,200.00
38-A-$2,000.00 38-D-$800.00 38-E-$400.00 39-F-$400.00
38-G - $275.00

Dalton

(Pictures 39-A thru 39-Z, 39-Z-1 thru 29-Z-6)

Dalton is a small to medium point. Well worked with an auriculate hafting area, usually strongly defined from the distal end. Dalton types average between 1-1/4 inches to 3 inches long and from 3/4 inch to 1-1/4 inches wide. Longer examples than 3 inches exist but are very rare (with the notable exception of flea markets and out of trunk dealers who seem to stock all early points in quantity).

The Dalton family has a number of traits that are shared by its variants. One such trait is the well-defined hafting area. There is a well defined change at the point where the blade ends and the haft begins. The blades are usually finely serrated or sometimes beveled. The haft is usually ground very heavily on the sides as well as the basal edge.

Dalton type points are found thinly distributed throughout the eastern U.S. where many local varieties are defined. The point appears on transitional Paleo sites throughout north Alabama and Tennessee, appearing to date about 10,000 years before present. Examples are found in association with Quad, Wheeler, Big Sandy I and biface and uniface tools. One common trait of this desirable point is the excellent workmanship which they exhibit.

Dalton variants are Breckinridge, Colbert, Greenbrier, Hardaway, Meserve, Hemphill, Nuckollls, Chatahoochie, Withlacooche and Chipola.

Withlacooche
39-A - $500.00

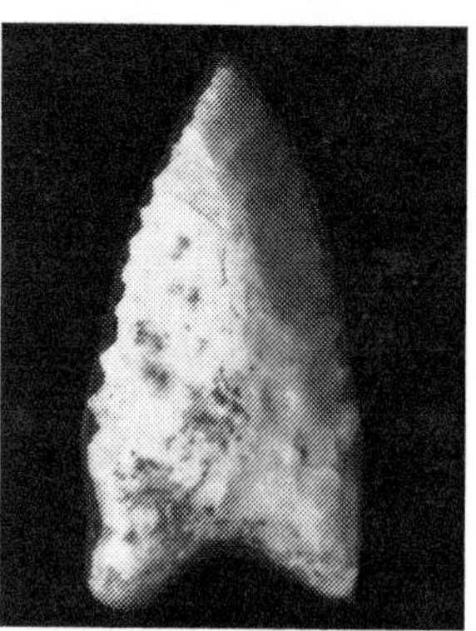

Withlacooche
39-W- $75.00

Withlacooche
39-B - $400.00

Withlacooche
39-C - $400.00

Withlacooche
39-D - $250.00

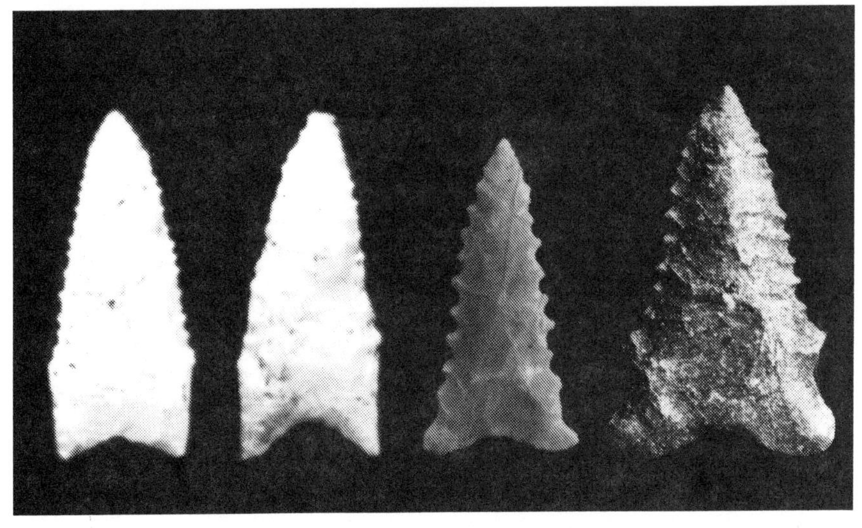

Chatahoochie
39-H - $75.

Chatahoochie
39-J - $35.
Tip Damaged

Hardaway
39-O-$100.

Hardaway
39-P-$175.

Nuckolls - 39-Q - $175.00

Breckenridge 39-F - $500.

Breckenridge 39-G - $275.

Hemphill 39-L - $350.

Hemphill 39-M - $175. *Tip Damaged*

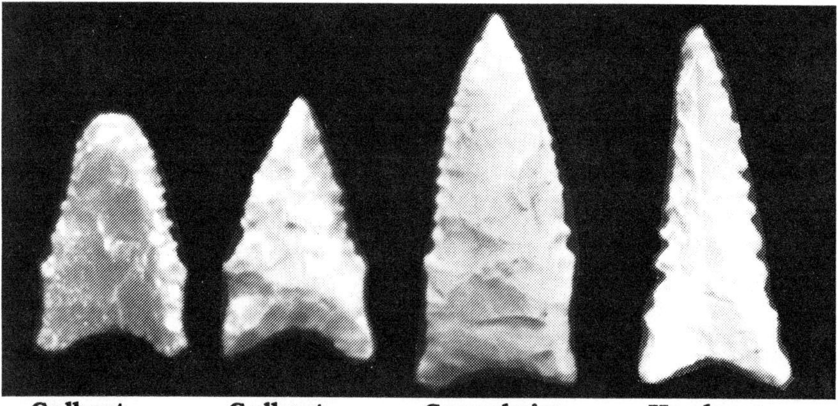

Colbert 39-X - $20. *Tip Damaged*

Colbert 39-Y- $80.

Greenbrier 39-Z-1- $65.

Hardaway 39-N - $120.

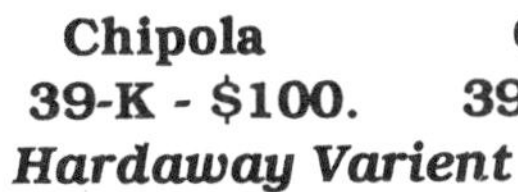

Chipola 39-K - $100. *Hardaway Varient*

Colbert 39-Z- $300.

Greenbrier 39-E- $175.

Nuckolls 39-R - $225.

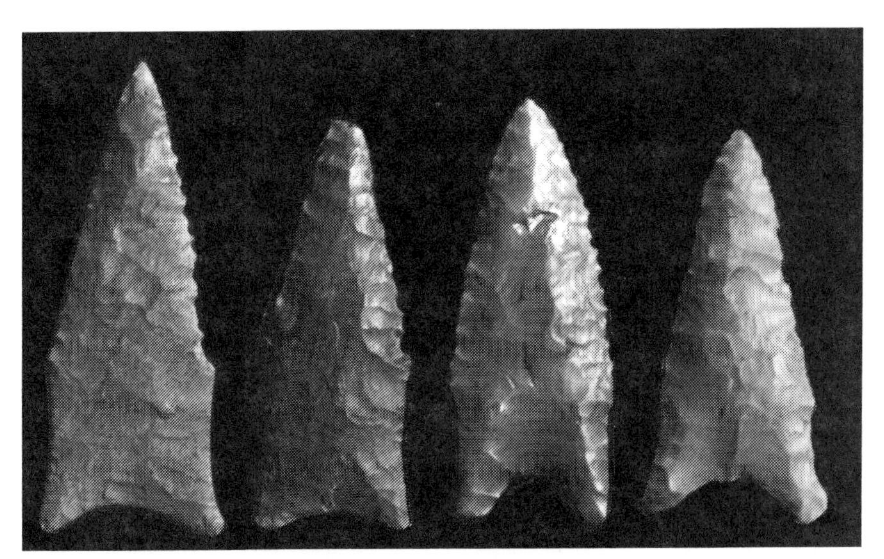

Greenbrier 39-T-$225.

Colbert 39-Z-2-$25.

Nuckolls 39-S- $200.

Greenbrier 39-U - $125.

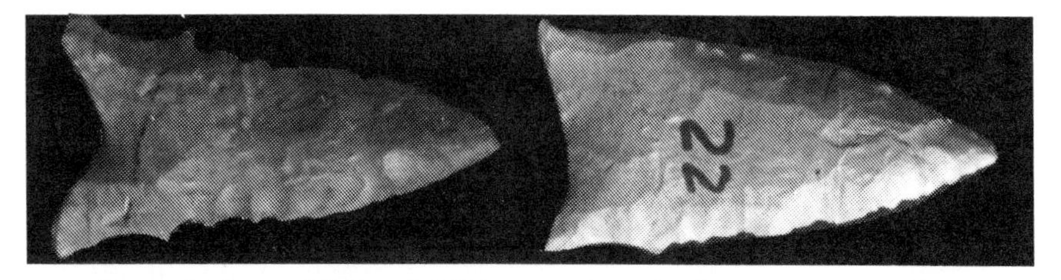

Greenbrier - 39-V - $125. **Nuckolls - 39-Z-3 - $75.**

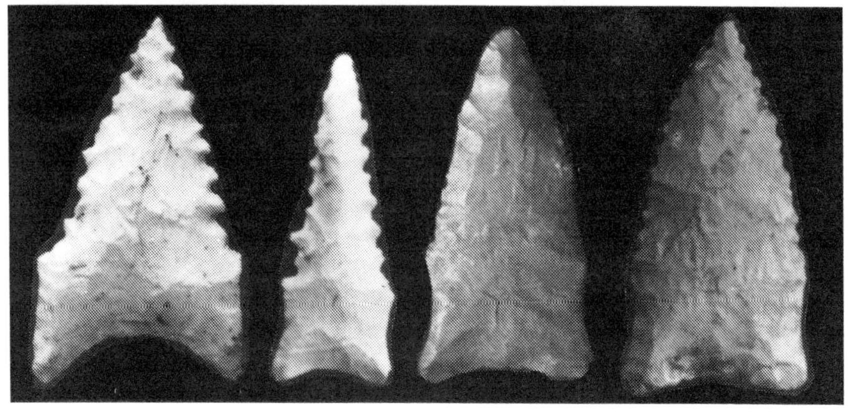

Chatachoochie 39-I - $250. **Colbert 39-Z-4 - $65.** **Colbert 39-Z-5 - $60.** **Colbert 39-Z-6 - $100.**

Damron

(Pictures 40-A thru 40-D)

Damron is a medium-sized point with small, shallow, side notches. This type measures 1-3/4 inches to 2-3/4 inches in length and 1-1/4 inches to a little over 1-1/2 inches in width.

Cross-section tends to be biconvex but may be flattened or plano-convex. Shoulders are tapered. The blade may be straight or excurvate, sometimes one side of each. The blade edges may be serrated. Distal end is acute. Hafting is defined by shallow side notches near the basal edge. The stem appears expanded with excurvate or straight basal edge that is rarely ground. Usually it is beveled on one or both faces.

Shaping of the blade utilized deep random flaking with fine retouch on blade edges. The retouch results in fine regular serrations on the blade edges. Notches are formed by removal of one or more flakes in the sides just above the basal edge. Short deep flakes resulted in the bevel of the basal edge. In cases where the basal edge wasn't beveled, they were thinned.

(continued on next page)

Named by Cambron after the Damron site in Lincoln County, Tennessee. The type was once known as Upper Valley Side Notched. Indications are of an early to middle Archiac association. Is thought to be a branch of the Benton culture.

40-A - $45.00 40-B - $20.00 40-C - $35.00 40-D - $35.00

Decatur

(Pictures 41-A thru 41-K)

A small to medium-sized point. It is a beveled, corner notched point with incurvate base often struck on basal edge. Examples range from less than 1 inch to over 2 inches in length and 3/4 inch to 1-1/2 inches wide.

Due to the steep bevel, the cross-section is nearly always rhomboid, only rarely biconvex. Shoulders are most often tapered with expanded barbs, but rarely straight, and either barbed or unbarbed. The blade will be straight or incurved. In rare cases blade may be recurvate. Blade edges are beveled on one side of each face and are often serrated. The distal is acute. The stem edges are expanded with straight edges. The basal edge is usually incurvate often having a flat place on the bottom of each side made by striking off a flake. The basal edge is usually thinned and often ground.

(continued on next page)

The blade and haft were formed by broad shallow flaking. Blade edges are serrated by retouch from the removal of deep flakes in the beveling process. Angle of bevel may be from steep to a more shallow bevel reaching half way across the face of the blade. One or more flakes were struck from the basal corners forming a flat area on the basal edge. Stem edges are then finished by grinding.

The type was named for the Decatur, Alabama area by Cambron. The point appears to be most concentrated along the Tennessee River Valley area. Examples are often found near this river on pre-shell mound sites. This leads to an early Archaic association.

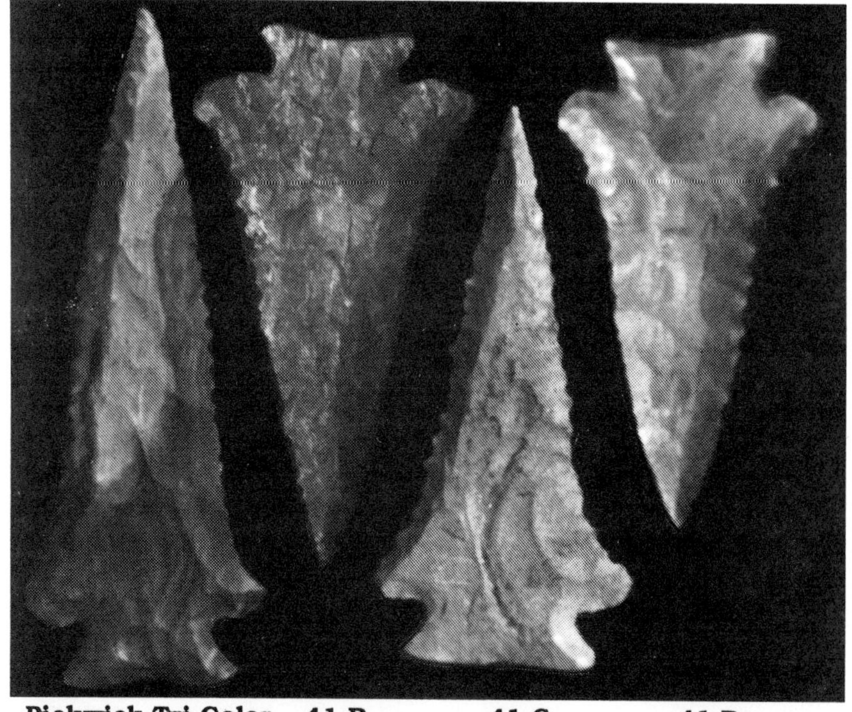

Pickwick Tri-Color 41-A -$2,500.00 **41-B $400.00** **41-C $500.00** **41-D $350.00**

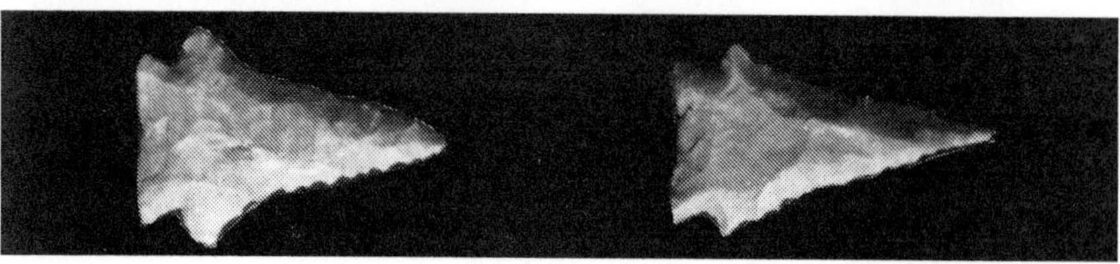

41-J - $60.00 **41-K - $35.00**

41-E	41-F	41-G	41-H	41-I
$200.00	**$275.00**	**$175.00**	**$75.00**	**$45.00**
	Rare Type			

Duval

(Picture 42-A)

Similar to the Mountain Fork and other "spike" points of the southeast. Has shallow side notches and a slightly expanding stem. Found on Woodland culture sites.

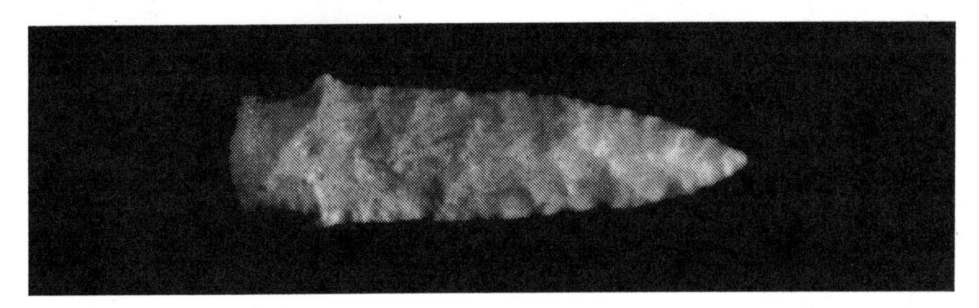

42-A - $75.00
Extra nice example. (1 inch average example, $5. to $35.00)

Ebenezer

(Pictures 43-A thru 43-F)

The Ebenezer is a small point with a short rounded stem and excurvate blade. The average length is around 1-1/2 inches and an average width of about 3/4 inch. The cross-section is biconvex.

Shoulders are narrow and either tapered or straight. Blade is most often excurvate with an acute point. The stem is rounded and short.

The blade and stem are formed by broad random flaking with short retouch flaking finishing the blade edges. This type was made from local material and often reflects the problems encountered from poor grade material.

The type was first called Rudimentary Stemmed when first recognized at the Camp Creek Site in Greene County, Tennessee. It was later named Ebenezer. It associates with the Woodland period. Assemblages seem to point to a date from around 2,000 years B.P. until around 1,500 years B.P. Note the type is similar to Clifton points from west of the Mississippi River.

43-A - $35.00 **43-B - $40.00** **43-C - $25.00**
43-D - $15.00 **43-E - $10.00** **43-F - $25.00**

Eccentric Notched

(Pictures 44-A thru 44-J)

This is a medium sized stemmed point. Blade edges have two or more large notches on each side of the blade.

This is a generic category and represents several cultures.

Periods range from Archaic to early Woodland.

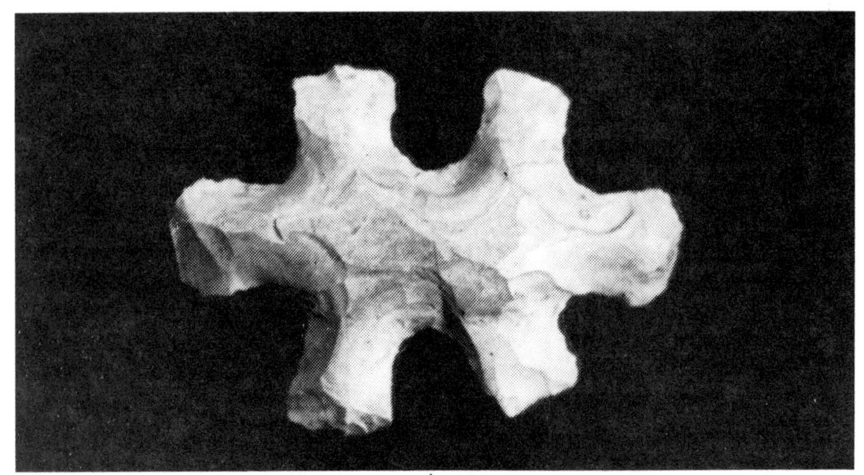

44-A - $450.00

44-B - $100.00 **44-C - $125.00**

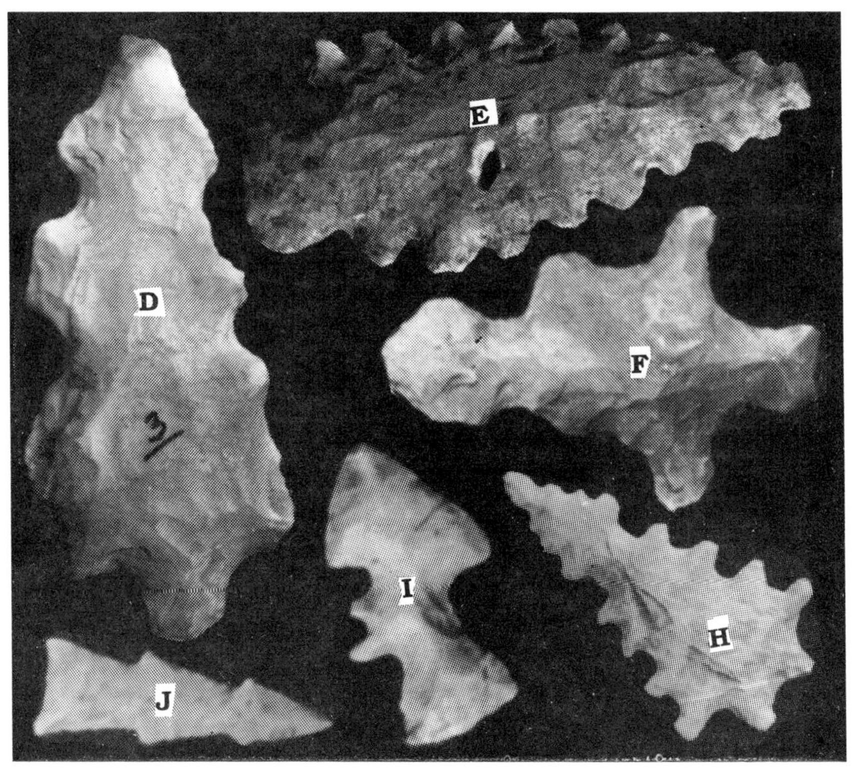

44-D - $100.00 **44-E - $100.00**
44-J-$175.00 44-I-$200.00 44-H-$300.00 44-F-$45.00

Ecusta

(Pictures 45-A thru 45-E)

Ecusta is a small beveled and serrated point, with a triangular shape and side notches. Lengths average about 1-1/2 inches and widths about 7/8 inches.

Cross-section is nearly always rhomboid. Beveling of the blade on one side of each face resulted in weak shoulder barbs. The blade is straight and usually serrated with an acute distal end. The hafting area is formed by notches in the side just above the basal edge. The basal areas is usually broad and rounded but in some cases may be straight or incurvate. Some examples have diagonally struck basal edges forming a flat area. This point was shaped by well controlled random flaking followed by retouching and serrating the blade with pressure flaking.

(continued on next page)

"Ecusta" is the Cherokee name for Davidson River location of the site in Transylvania County, North Carolina, where the type was first recognized by Harwood in 1958. These points share a number of characteristics of the Plevena and Decatur such as beveled blade, side notches and struck bases. Always being found with early Archaic artifacts indicates an age of over 5,000 years.

45-A	45-B	45-C	45-D	45-E
$45.00	**$35.00**	**$30.00**	**$15.00**	**$3.00**

Elk River

(Pictures 46-A thru 46-G)

This type is a medium to large stemmed point. Points are always transverse-oblique flaked. Average length from less than 2 inches to over 4 inches. Longer examples exist but are quite rare. The average width is from 1-1/8 inches to 1-1/2 inches.

Occasionally the cross-section may be plano-convex but is almost always biconvex. The blade is excurvate with acute distal end. The hafting area is generally straight stemmed with straight basal edge and stem sides. Some have ground bases with most being unground.

The faces were generally both shaped by transverse-oblique flaking. The blade and stem were then finished by fine retouch. Very little retouch was required due to the characteristics of oblique flaking. There is in most cases a median ridge.

Elk River was named for sites on the Elk River of Limestone County, Alabama. Most sites in Alabama and Tennessee are from the shell mound period, indicating an age of around 5,000 years before present. This would place the type in middle to late Archaic period.

(points on following page)

46-A-$800.00　　46-B-$350.00　　46-C-$275.00

46-D-$75.00

46-E-$75.00 46-F-$60.00 46-G-$50.00

Elora

(Pictures 47-A thru 47-C)

Elora is a thick, medium to large point with an unfinished stemmed base. Length ranges from 2 inches to 3 inches and a width from 1-1/2 inches to over 2 inches.

Cross-section is biconvex. Shoulders are usually tapered and rounded, but may be horizontal or, in rare instances, expanded. Blade edges are usually straight, some are finely serrated by retouch. The distal end is acute or rarely broad. The stem is constructed with incurvate or straight sides. The basal edge usually has an unfinished or broken look, this results from removal of one or more basal flat flakes across the bottom.

The blade and haft were formed with broad random flakes that may be shallow or deep. The retouching, by taking flakes from alternating sides, resulted in fine serrations on some examples. Some secondary flaking was used to form the shoulders and sides of the stem. This retouch around the base was often very steep. The basal edge, sometimes called "fracture base", may either be unfinished or purposely flaked in that manner.

The type was named for the Elora area of Lincoln County, Tennessee, where numerous points of this type have been found. The type was named by Cambron in 1960. A local variant is called Elora Serrated because of the strong serrations found when deep pressure flakes were removed from alternating faces. The Elora type was once called Provisional Type 7 by Cambron. Evidence points to middle to late Archaic to possibly early Woodland association. Possibly 5,000 to 3,000 years before present.

47-A - $35.00 **47-B - $20.00** **47-C - $15.00**

Eva

(Pictures 48-A thru 48-E)

Eva is a medium to large excurvate or recurvate point with basal notches. Eva ranges from under 2 inches to over 4 inches in length and 1 inch to 1-1/2 inches in width.

The cross-section is biconvex. The blade outline may either be excurvate or recurvate. Shoulders are inversely tapered, simple barbs. Basal edge has two notches averaging a little over 1/8 inch deep. The stem in the center is thus formed. The stem is either straight or slightly contracted. The stems basal edge is usually straight and thinned by fine retouch. The stem and/or notches are often ground.

Shaping of the blade and haft area is accomplished with broad, strong, random flaking. Blade faces are thinned from edge to center by broad secondary flaking. The blade as well as the hafting area will show some fine retouch to the edges. The basal notches are formed by the removal of one broad flake from each face on each side of the stem. Barbs are often shorter than the stem due to retouch on them. No retouch is usually evident in the notches.

The point is named for the Eva site in Benton County Tennessee and was first noted by Kneberg in 1956. The type was associated with other types that were carbon dated at 7,200 years before present, at the Eva site. A great deal of variation has been noted in this type, especially in the shape of the blade and the width of this point type. Numerous finds in north Alabama seem to agree with the Archiac date of around 7,200 B.P.

48-A - $1,500.00

48-B-$200.00 **48-C-$1,200.00** **48-D-$300.00** **48-E-$45.00**
Pickwick Tri-Color **This point was stolen, contact author if seen!**

Evans

(Pictures 49-A thru 49-C)

Evans is a medium sized point with an unusual notching in the blade. Usually they have expanded stems. The type averages around 2-1/4 inches in length and around 1-1/4 inches in width.

Cross-section is biconvex. Shoulders may either be tapered, inversely tapered, or horizontal. The blade is excurvate or sometimes straight. The distal end may be acute or broad. There are two notches taken from the blade about 1/3 to 1/2 way between the shoulders and the distal end. The stem is generally expanded but in some points may be straight or even contracted. The basal edge is straight or sometimes incurvate, and is sometimes ground.

(continued on next page)

Broad, irregular, random flaking was used to form the blade and haft. Blade edges show small, deep, flaked retouch. Deep side notches are formed by removal of one flake from the edge of opposite faces, sometimes with a little fine retouch to finish their shaping.

The type was named after examples found at Poverty Point site in Louisiana. Distribution of the points cover Louisiana, east Texas, southern Arkansas and western Mississippi. Dates of around 2,500 to 3,000 years seem indicated for this type.

49-A-$75.00 49-B-$35.00 49-C-$50.00

Flint Creek

(Pictures 50-A thru 50-F)

Flint Creek points are medium to large, serrated, stemmed and often relatively thick. The type averages from 1-1/2 inches to 3 inches in length and 3/4 inch to 1-1/4 inches in width.

Cross-section is always biconvex. The blade is excurvate and usually finely serrated. The shoulders may be inversely tapered, tapered, or sometimes horizontal. The distal end is acute. The stem is expanded, with excurvate side edges or rarely straight side edges. The basal edge of the stem may be unfinished and excurvate and is often lightly ground.

(continued on next page)

The blade is shaped by broad random flaking. The thick blade was pressure flaked with deep, narrow, sometimes long flakes, often resulting in fine serrations along the blade edge. These flakes were removed by alternating the faces as chipping progressed along the blade edges. Broad deep flakes taken from the corners or sometimes sides, shaped the stem. Except in the straight stem type, the basal notches were usually unretouched. Basal edges sometimes shows broad shallow thinning flakes.

The point was named for Flint Creek in Morgan County, Alabama by Cambron. A peak period for this type seems to have been in late Archaic to early Woodland periods.

50-A-$150.00 **50-B-$35.00** **50-C-$45.00** **50-D-$10.00**
(Actual Size- 3-1/2") **50-E-$10.00** **50-F-$5.00**

Flint River Spike

(Pictures 51-A thru 51-E)

This is a small to medium narrow spike shaped point. The type averages from 1-1/2 inches to 2-1/4 inches in length and from less than 1/2 inch to about 3/4 inch in width.

Cross-section may either be biconvex or median ridged. The base will generally be rounded. Some examples may have an unfinished basal edge. Blade edges are most often excurvate but may be straight sided with acute distal end. The hafting area includes areas between the basal edge and the widest point of the blade. Base is often thinned. Random percussion flaking shaped the blade and haft. The flake scars are often rather deep, with about 1/2 examples showing retouch on blade edges.

The type derives its name from Flint River Mound at the mouth of Flint River in Morgan County, Alabama. Flint River Spike and Bradly Spike appear together on sites and may be related types. The Flint River has no stem whereas the Bradly is distinctly stemmed. The time frame for appearance of this type appears to be late Woodland.

51-A $35.00 **51-B-$5.00**
51-C-$5.00 **51-D-$10.00** **51-E-$5.00**

Fort Ancient

Fort Ancient is a small to medium rather thick point, often serrated. The average length is around 1 inch to almost 2 inches and averaging about 1/2 inch in width.

The type has a biconvex cross-section. The blade is usually straight but may be slightly excurvate. The distal end will be either keen and acute or acuminate. The base may be straight or excurvate, usually thinned and expanded. Serrated examples are more easily identified as classic examples are noticeably serrated.

The point type is shaped by random rather deep flakes and generally finely retouched near the distal end. The serrations result from removal of flakes from alternating faces. The serration is usually more irregular than in most serrated types. The basal edge was usually retouched and thinned by removal of broad flakes.

Fort Ancient points are named for the Fuert Focus of the Fort Ancient aspect of the Ohio Valley, where they were first recognized. It is found on sites with Madison, Jacks Reef Corner Notched, Knight Island and related points from the Woodland and Mississippian culture. The type has been found with European trade items suggesting a date as late as the 1600's.

A............................$35.00
B............................$45.00
C............................$50.00
D..........................$100.00
E............................$75.00
F..........................$150.00
G............................$15.00
H..............................$5.00
I...............................$5.00
J..........................$600.00
K............................$35.00
L............................$10.00
M............................$10.00
N..........................$250.00
O..........................$150.00
P..........................$100.00
Q............................$75.00
R............................$10.00
S............................$10.00

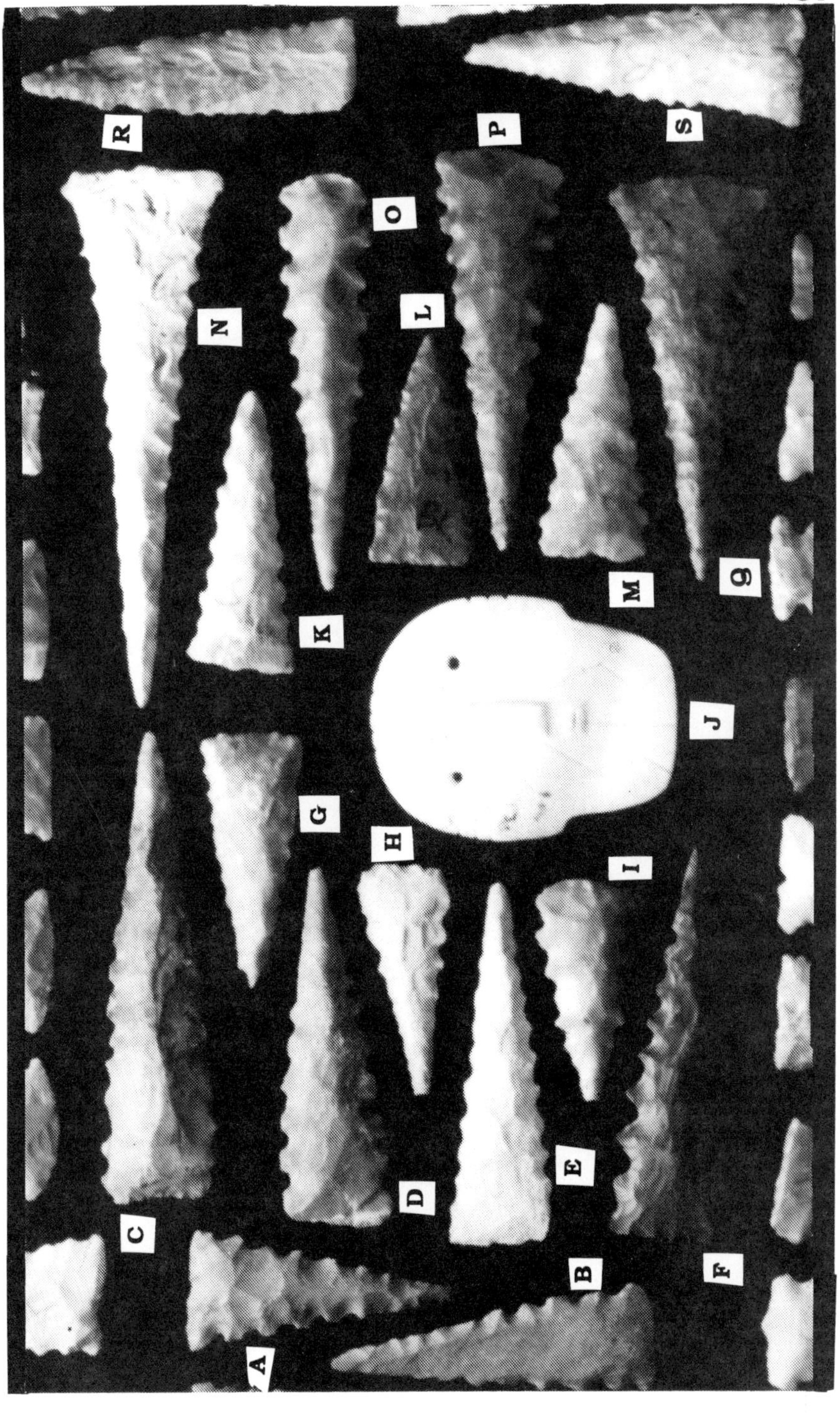

Frazier

(Pictures 53-A thru 53-E)

Frazier is a medium sized, triangular point. The point will have a well thinned basal edge. Average lengths from 2 inches to 2-3/4 inches and averages 3/4 inch to 1 inch in width.

Cross-section is flattened. The distal end is acute. The blade is excurvate then becomes parallel in the undefined hafting area just about the basal edge. Blade edges may appear slightly serrated. Basal edge is usually straight but may be slightly incurvate. The point may rarely be ground but will always be well thinned. Shaping of the blade is done with large shallow flakes, followed by shorter, deeper flaking along the edges. Some retouch may result in a sort of serration on the blade edges. The basal edge is thinned by broad fairly long flakes.

The type is named from the Frazier site in Benton County, Tennessee by Kneberg. No examples have been found in context, undisturbed, so dating is only conjectural by association on surface sites. Estimated age is around 1,500 B.C. to early centuries A.D. Examples in Tennessee Valley are found with early Archaic point types. This type is similar to Paint Rock Valley points but is narrower with less incurved basal edge, and better worked.

53-A-$35.00 **53-C-$30.00**
53-D-$15.00 **53-E-$15.00**

Garth Slough

(Pictures 54-A thru 54-F)

Garth Slough is a small to mid-sized point with expanded barbs. The barbs are usually obtusely terminated. Lengths range from barely over 1 inch to about 1-7/8 inches and around 1 inch to 1-1/4 inches wide.

The type has a biconvex cross-section. The blade edges are usually incurvate, or rarely, recurvate or straight, with acute distal end. The shoulders are expanded barbs and are usually broad. The ends of the barbs are either obtuse or straight. Most examples are finely serrated. The stem area is formed by removal of diagonal notches. The basal edge of the stem is usually straight or excurvate, thinned and may be ground.

The broad random flaking used to shape the faces was usually nearly obscured when the secondary flaking along blade and base was executed. This secondary flaking in some points resulted in a slight median ridge and fine serration on the blade edge. Broad deep flakes were removed to form the basal notches.

The type is named for surface finds on sites around Garth Slough in Morgan County, Alabama, and was named by Cambron. The type is found on transitional Paleo and early Archaic sites in north Alabama in conjunction with Big Sandy, Dalton, Stanfield, Lerma, and Jude .

54-A - $250.00 **54-B - $175.00** **54-C - $100.00**
54-D - $75.00 **54-E - $45.00** **54-F - $15.00**

Gary

(Pictures 55-A thru 55-D)

Gary is a medium size point with a contracted stem. Average length is from 1-1/2 inches to 3-1/8 inches and around 1 inch in width.

Cross-section is usually biconvex. Shoulders are tapered or horizontal. In some examples the shoulders may be rounded or expanded barbed. The blade is generally straight to excurvate but in some examples may be incurvate or recurvate. The type has an acute distal end. The basal edge of the stem may be rounded to slightly pointed, with straight or excurvate stem edges. The stem is usually contracted.

Broad random flaking was used to shape the blade and stem. The edges of the blade show short deep retouching flakes. There may be some retouching around the stem edges.

The type was named after the Gary Contracting stem type, a Texas point type. Excavated examples as well as surface finds associations seem to indicate late Archaic to early Woodland period culture. Note: There is a number of variations in this type, attention to flaking may help in some cases to identify this type.

55-A - $45.00

55-B - $35.00

55-C - $75.00
55-D - $35.00

Greenbrier

(Pictures 56-A thru 56-O)

Greenbrier is a medium to large point usually very well worked, with expanded auricles and vague to strong side notches. This type averages over 1-1/2 inches to 2-1/2 inches in length. Rare examples 3 inches and over exist but are very seldom seen. Average width is between 1 inch and 1-1/2 inches.

Cross-section may be biconvex but is usually flattened. Shoulders are the result of the shallow side notches. The shoulders will be tapered or very weakly barbed. The blade edges are usually parallel with acute distal end. The blade is usually finely serrated and in some points may be beveled on each side of both faces. The haft is a sort of expanded auriculate stem formed by the removal of side notches. The basal edge is incurvate and thinned. The sides and bottom of haft is heavily ground on most examples. Blade shape and base was formed by broad thin flakes. Blade edges and sometimes haft sides were beveled with short steep flakes. Notches were made by the removal of one or two large flakes then finely retouched and usually heavily ground. Usually, material was Ft. Payne chert or another high grade type. The Greenbrier is often an example of the highest point of the flaking art.

(continued on next page)

The type was described by Lewis and Kneberg in 1960. Greenbrier is often found on surface sites with Pine Tree points and may somehow be related. In general the type is thinly distributed in the southeast with many fine examples occurring in north Alabama and south Tennessee. A likely age would be from 5,000 to 9,000 years B.P., early Archaic to transitional Paleo periods.

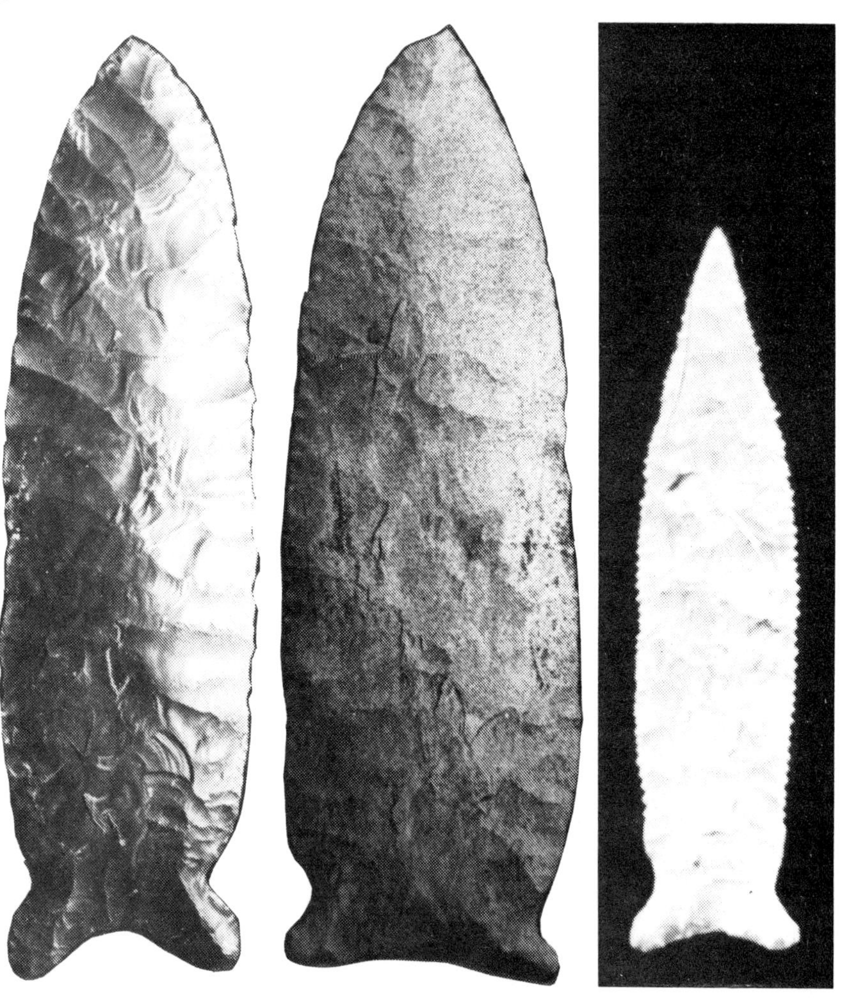

56-A - $1,200.00 **56-B - $850.00** **56-C - $1,500.00**

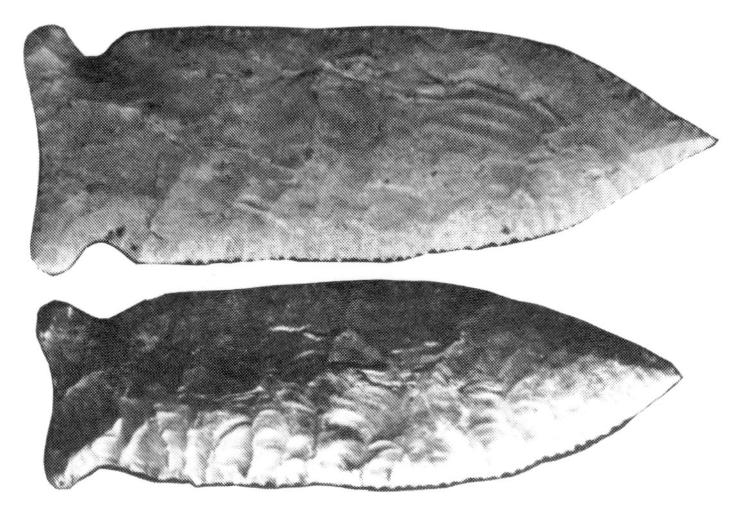

56-D - $1,800.00
56-E - $750.00

56-F- $500.00 56-G - $450.00 56-I - $900.00 56-J - $750.00
Pickwick
Tri-Color

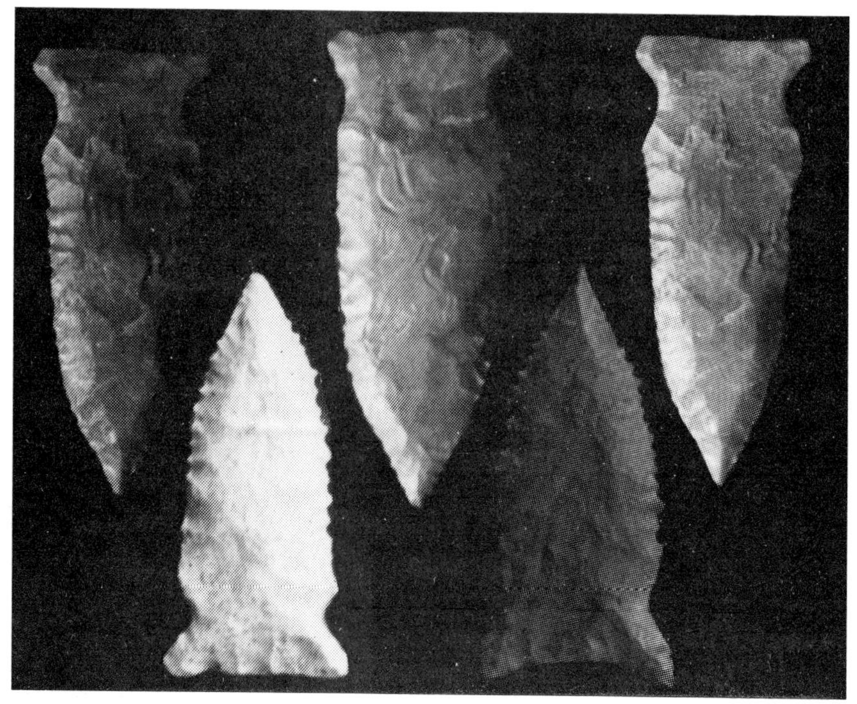

56-K - $100.00 **56-L - $125.00** **56-M - $100.00**
56-N - $90.00 **56-O - $125.00**

Greenville

(Pictures 57-A thru 57-E)

Greenville is a small point with a triangular shape and parallel to excurvate basal edges. Length averages from 1-1/4 inches to 2-1/2 inches long and 3/4 inch to 1 inch in width.

Cross-section is biconvex. Shape of the blade is excurvate or straight above the haft. The hafting area sides will be parallel or taper slightly toward the distal end of point. Basal edge is straight or slightly incurvate and is usually thinned.

Random flaking shaped the blade and haft area. Usually some secondary flaking will show along the blade edge. This retouch is usually fairly deep and narrow to broad. Broad flakes were taken from the basal edge to thin it and was often followed up with some retouch. Material varied with local availability.

(continued on next page)

The type was named for the Camp Creek site near Greeneville, Tennessee. Excavations have revealed a date around 2,000 years before present. This type is spread along the western edge of the Appalachians from Greeneville, Tennessee to south Alabama and along the Tennessee River Valley. The type was named by Lewis and Kneberg in 1957.

57-A - $150.00 **57-B - $25.00** **57-C - $10.00**
57-D - $10.00 **57-E - $5.00**

Guilford

(Pictures 58-A thru 58-D)

Guilford is a medium to large oval point with an incurvate base. This type ranged from under 2 inches to nearly 5 inches in length and 3/4 inch to 1-1/2 inches at the widest point.

The cross-section is usually rather thick and biconvex or sometimes slightly median ridged. The distal end is either acute or apiculate. Blade edges are usually excurvate but may be nearly straight. The haft area has contracted sides with rounded short auricles and an incurvate base. Sides of the haft may be lightly ground but may otherwise be hard to define between haft area and blade edge.

(continued on next page)

Well controlled random flaking sloped the blade and haft area, flaking rarely may approach transverse-oblique. The side edges were retouched with uniform, short and often deep flaking. The basal edge was formed by removal of one or more broad flakes from each face. The type was made from a wide variety of local materials.

The type was named for the Guilford focus of the Carolina Piedmont and was first noted by Coe in 1952. An association of over 5,000 years old, of the early Archaic period. Points A & B are Guilford. Points C & D are Guilford Round Base.

58-A - $75.00
58-B - $35.00
58-C - $75.00
58-D - $10.00

Guntersville

(Pictures 60-A thru 60-H)

Guntersville is small to medium point, with excurvate blade and straight basal edge. The type may range from around 1 inch to about 2 inches in length and 3/8 inch to 5/8 inch in width at widest pint of the blade.

Cross-section may be biconvex but is usually flattened. The widest point of the excurvate blade may be just below half-way from basal edge or at the basal edge. Side edges of the hafting area may be nearly parallel to slightly contracting. The basal edge is thinned and straight.

Shaping of the blade and haft was performed by removal of broad, shallow, random flakes followed by fine retouch on the entire blade and the haft edges. Elongated shallow flakes were removed from the basal edge to thin the base. Materials used varied subject to local availability.

Guntersville was named for the Guntersville Basin of the Tennessee River, by Cambron. The type seems to have appeared during the late Mississippian period and survived to early historic contact. The type has been connected to historic Cherokee sites. Kneberg states a possible age of 1,300 A.D. until as late as 1,800 A.D. In Florida this type is called Ilchetuchnee.

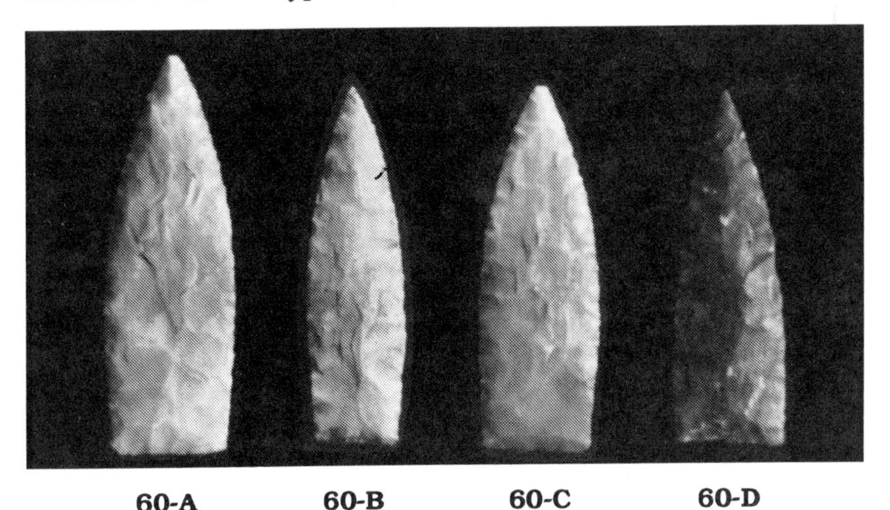

60-A	60-B	60-C	60-D
$125.00	**$75.00**	**$30.00**	**$35.00**
		Tip Damaged	

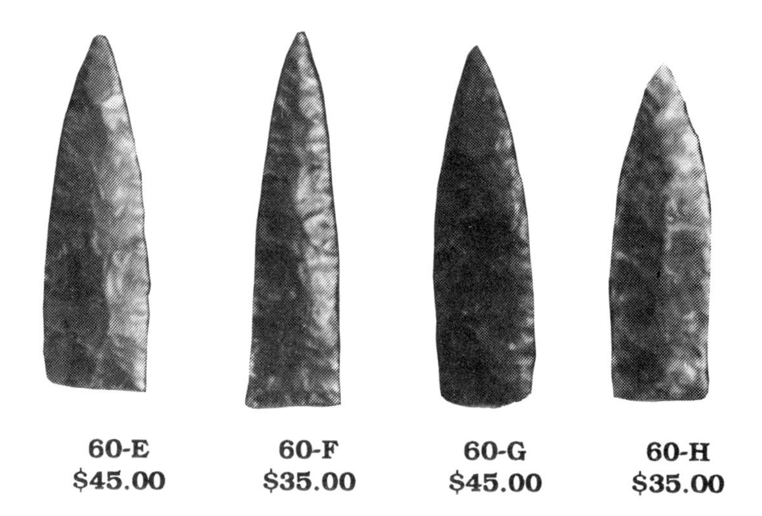

60-E **$45.00** | 60-F **$35.00** | 60-G **$45.00** | 60-H **$35.00**

Halifax

(Picture 61-A)

Halifax is usually made of vein quartz or quartzite. The point is small to medium with side notches. Length ranges 1 inch to 2-1/2 inches on average and 5/8 inch to 1 inch wide at shoulders.

Cross-section is biconvex. The blade edges are usually straight or occasionally excurvate. The shoulders are usually tapered. The stem sides will be straight or incurvate with a straight or excurvate basal edge. The stem will be expanded, the distal end will be acute. Basal edge and side notches are usually ground. Shaping of the blade and haft area employed broad, usually deep, random flaking along with good secondary flaking on some points. The shallow side notches are formed by removing several deep flakes. Examples are often rather thick, often being chipped from cores taken from quartz or quartzite boulders.

The name is derived from Halifax County, North Carolina, being named such by Coe in 1959. Halifax culture in North Carolina may be related to Swan Lake in the Tennessee River Valley area. Suggested age would be about 5,000 years B.P. or a late Archaic to early Woodland association.

61-A
$10.00

Hamilton

(Pictures 62-A thru 62-G)

Hamilton is a small, triangular point with incurvate base and incurvate basal edge. The type will average from less than 3/4 inch to near 1 inch in length and averaging 5/8 inch in width at points of basal edge.

Hamilton will have a flattened cross-section. Blade is usually incurvate but may be nearly straight. The distal end will be acute. The base is generally incurvate but may be nearly straight. The basal edge is always thinned. The delicate chipping of these finely made points required skilled pressure flaking. Basic shaping employed broad shallow flaking finished around the edges by extremely fine pressure retouch. Broad random flakes were removed from basal edge to thin the base. Occasionally one may see a little retouch near the tips of basal edge but rarely any retouch toward middle of base.

Hamilton type was named for sites in Hamilton County, Tennessee, where it was first recognized by Lewis in 1955. This type is found relative to the large dome-shaped burial mound cultures. Examples appear to be very late Archaic through the mid-Woodland.

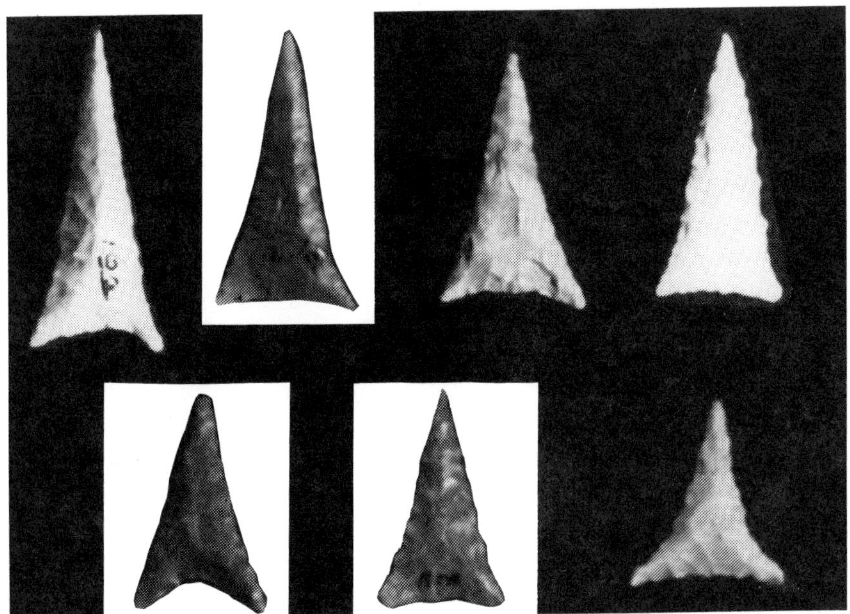

62-A - $35.00 62-B - $30.00 62-C - $10.00 62-D - $30.00
62-E - $15.00 62-F - $10.00 62-G - $10.00

Hamilton, Stemmed

(Pictures 63-A thru 63-D)

Hamilton Stemmed is a medium-sized, excurvate bladed, expanded stem point. Average for the type is around 2-7/8 inches in length and 1-1/8 inches in width.

Hamilton Stemmed has a biconvex cross-section. The blade is excurvate with an extremely acute distal end. Shoulders are inversely tapered terminating in sharp barbs. The stem is expanded usually with straight stem sides. The basal edge is thinned and either straight or somewhat excurvate.

The faces of the blade and haft area were shaped by broad, shallow to deep random flaking. Retouching may be present on one or more edges on different examples. Corner-notches are the result of removal of strong broad flakes at the corners of the blade. Thin stem edges often were retouched along sides and fine flakes thinned the base.

Type gets its name from Hamilton County, Tennessee where the Hamilton culture was first recognized by Cambron. This type is found in association with Woodland period domed mound cultures. This association seems to indicate an early Woodland time period.

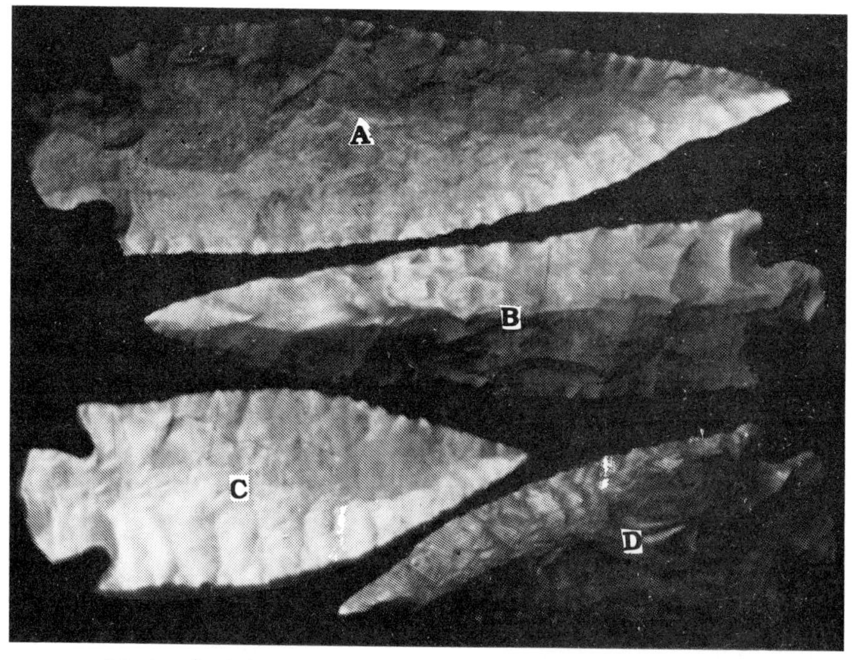

63-A - $125.00 **63-B - $75.00**
63-C - $150.00 **63-D - $45.00**

Hardaway

(Pictures 64-A thru 64-B)

Hardaway is a small to medium point with incurvate base and side notches. Average lengths range between 1 inch and 2 inches long and are from 3/4 inch to 1 inch wide.

Hardaway will have a biconvex cross-section. The blade is usually straight but in some examples may be excurvate. Blade edges may be serrated and point is acute. The hafting area is side notched with an auriculate base with rounded expanded auricles and concave basal edge. The basal edges is thinned and nearly always ground in notches and basal concavity.

Broad, well controlled flaking shaped the blade and haft. Often when looking at the point of the distal end one will note a slight median ridge formed by the fine secondary flaking that meets in the middle of the face. The basal edge was thinned by removal of broad shallow flakes. The entire edges of both blade and haft were finely retouched.

The type was named after the Hardaway site in Stanley County, North Carolina. The type was named by Coe. A radio carbon date around 9,000 to 1,000 B.P. indicates a transitional Paleo association.

64-A - $250.00 **64-B - $75.00**

Hardee

(Pictures 65-A thru 65-B)

Hardee is a thick stemmed point found along the Gulf Coast of America.

Usually it is alternately beveled and may be serrated.

Stem edges are often ground.

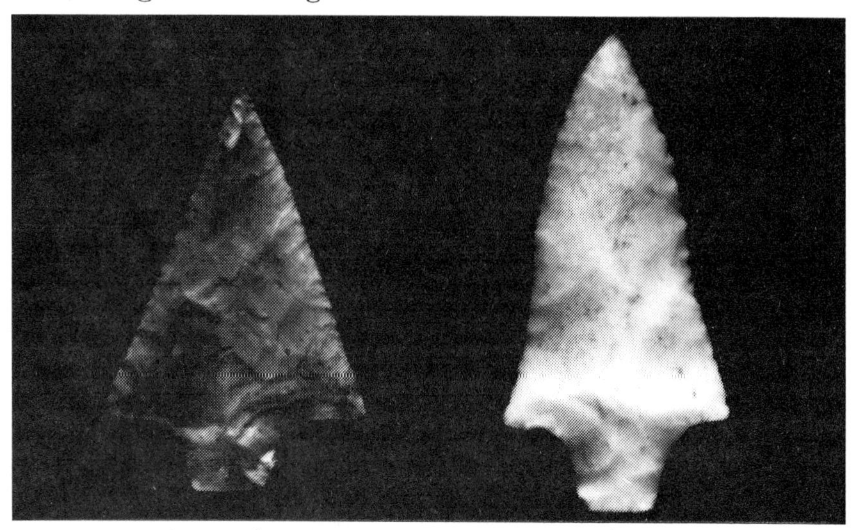

65-A - $275.00 **65-B - $200.00**

Harpeth River

(Pictures 66-A thru 66-I)

Harpeth River is a medium to large, side notched point with beveled blade edges and a stemmed base. The type averages from 2 inches to about 3-1/2 inches in length and 3/4 inch to 1 inch at the shoulders.

Cross-section will be flattened. The blade may be excurvate or straight and beveled on each side of both faces. The distal end is usually acute. Shoulders are tapered and haft sides are incurvate and ground. The basal edge is straight, nearly always ground, and is thinned. Shallow side notches are removed making an expanding stem. The blade and hafting area were formed by broad shallow random flaking. Both edges of each face are retouched by removal of deep, short, narrow flakes along blade edges resulting in fairly steep bevels on both sides of each face and a sort of fine serration.

(continued on next page)

The shallow notches also are steeply retouched often resulting in expanded shoulder barbs. This type is generally very well worked and of good material.

This point was named from sites along the Harpeth River in Cheatham and Dickson Counties in Tennessee. Examples were found on sites with such types as Big Sandy, Dalton, Greenbrier, Pine Tree, Cumberland, LeCroy, and Copena. Note that in some examples, the base area may resemble Dalton, with it's stem approaching auriculate. Workmanship may be very similar to types such as Pine Tree. An early Archaic back to transitional Paleo time span is suggested.

Editor's Note: A cache of points was found in January 1993, which contained four Kirk stemmed points and one classic Harpeth River point.

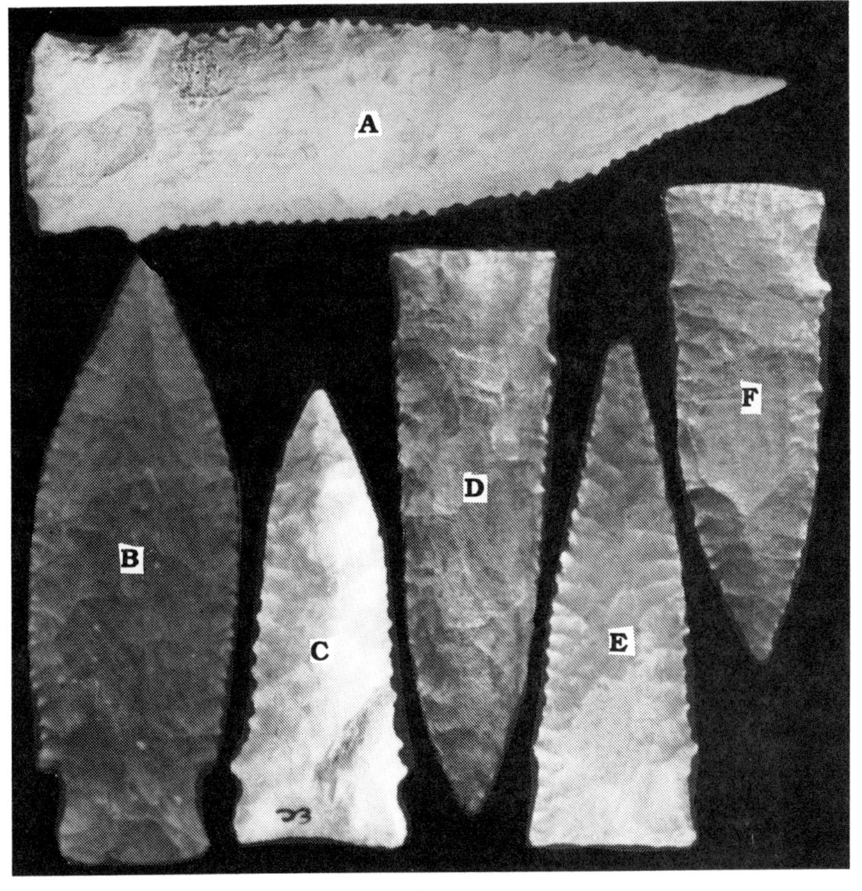

66-A - $1,500.00

66-B - $800.00 **66-D - $275.00** **66-F - $200.00**

66-C - $175.00 **66-E - $225.00**

66-G
$300.00

66-H
$350.00

66-I
$125.00

Hernando

(Pictures 67-A thru 67-D)

A small to medium basal notched point.

Blade edges may be straight or serrated.

Knapping techniques are similar to the Santa Fe or other Gulf formational type.

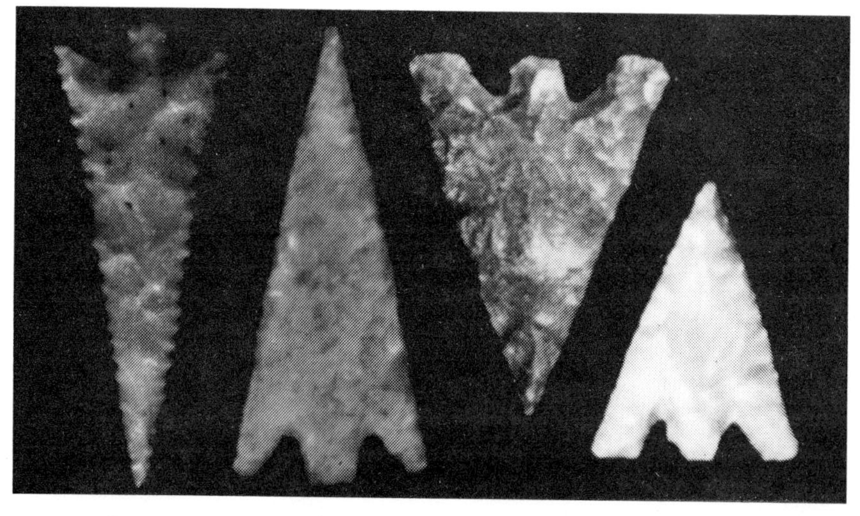

67-A-$450.00 67-B-$225.00 67-C-$175.00 67-D-$35.00

Hillsborough

(Pictures 68-A thru 68-C)

The smallest of the Newnan cluster; also called Batwing point by Florida collectors. The base and shoulders are barbed or pointed.

68-A - $1,500.00

68-B - $1,200.00 **68-C - $800.00**

Jacks Reef Corner Notched

(Pictures 69-A thru 69-F)

This type is a medium sized, corner notched point that is usually thin and flattened. Average length is around 1 inch to 2 inch and 1/2 inch to 1 inch in width.

Cross-section is usually flattened but may in rare cases be plano-convex. The blade may be either excurvate or parallel angular. The distal end may be from narrow to broad and acute. Shoulders are tapered with sharp, pointed, thin barbs. The hafting area is defined by the deep diagonal notches into the corners of the base. The base is an expanded stem with straight or rarely incurvate basal edge. Base is thinned and may be lightly ground.

Shaping of the blade and haft area utilized broad, shallow, random flaking followed by well controlled, broad, shallow secondary flaking. The blade edges were then finely retouched. Broad flakes were stuck from the corners to form the hafting area notches. The notches were then finely retouched. Material varies greatly.

The point was named from the Jacks Reef Site in Onondaga, New York, where it was first noted by Ritchie. Radio carbon dates the type in Alabama at about 1,500 years before present, placing its occurrence in Woodland period.

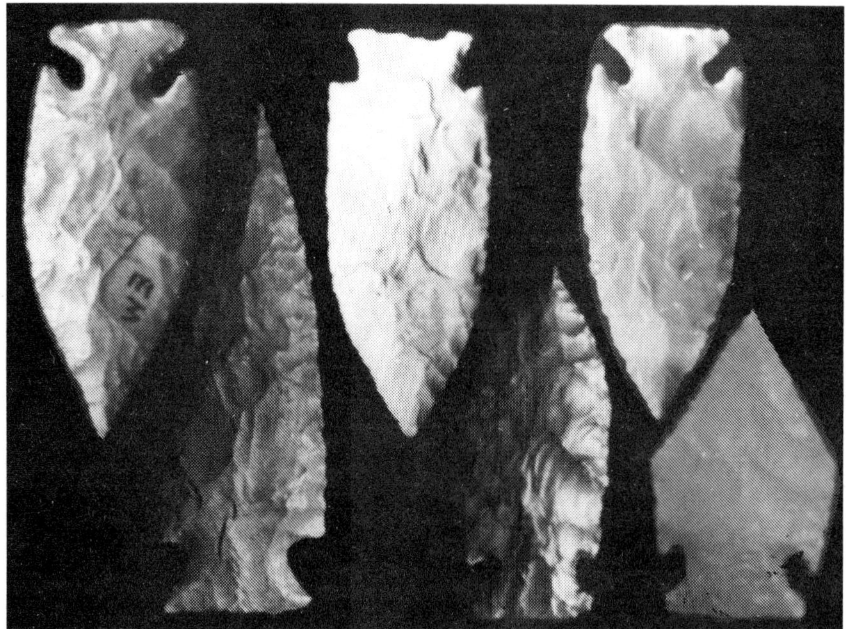

69-A- $200.00 **69-B - $175.00** **69-C - $45.00** **69-D - $125.00** **69-E - $75.00** **69-F - $150.00**

Jacks Reef Pentagonal

(Pictures 70-A thru 70-E)

This point is small to medium, pentagonal and usually quite thin. Length averages from 1 inch to 1-1/2 inches and 1/2 inch to 1-1/4 inches at the widest point.

This type will have a flattened cross-section. The blade edges are straight with an acute point. The hafting area is contracted with straight sides. Hafting area may be 1/2 to 1/3 the length of the point. Basal edge may be straight or slightly incurvate and is thinned. Point where haft and blade meet is usually well defined.

Broad, shallow, random flaking was used to shape blade and haft area. Fine retouch will be seen on all edges. Made from local materials.

Type was named after the late Point Peninsula Jacks Reef site in Onondaga County, New York. Workmanship and association appear to place this type between the Copena Triangular and the Madison point. A probable date of around 1,000 years old or older seems reasonable. This indicates a late Woodland to early Mississippian occurrence.

70-A - $35.00 **70-B - $125.00** **70-C - $45.00**
70-D - $7.00 **70-E - $10.00**

Jeff

(Pictures 71-A thru 71-D)

Jeff is a medium size broad point with expanded auricles and a straight base. Lengths range from 1-1/8 inches to 2-3/4 inches and 1 inch to 1-1/4 inches in width.

Jeff has a flattened cross-section. Blade shape is excurvate, sometimes slightly beveled on one side of each face. The blade edge may be finely serrated. The hafting area is expanded rounded and the distal end is acute. The basal edge is most often straight but may be slightly incurvate and is usually thinned or beveled. Hafting area sides are defined by grinding on the edges.

The blade was shaped and flattened using broad thin fairly long flakes. Secondary flaking along the edges are shorter and deeper and may result in fine serrations on the blade edges. Short retouching was applied to sides of the hafting area. Retouch on basal edge, due to steepness, gives the base a bevel.

The type was named after the Jeff area of Madison County, Alabama where it was first recognized. The type is always found with uniface tools and other pre-shell mound artifacts. This evidence points to a transitional Paleo association.

71-D
$75.00

71-A **$45.00** | **71-B** **$45.00** | **71-C** **$35.00**

Jude

(Pictures 72-A thru 72-G)

Jude is a small stemmed point with a short blade. Average length is 3/4 inch to just over 1 inch in length and from 1/2 inch to 1 inch in width.

The cross-section is biconvex or occasionally plano-convex. Blade edges are straight only rarely excurvate. The distal end is acute. Shoulders are usually horizontal but may be tapered or inversely tapered. The stem is straight or sometimes slightly expanded. Basal edge is slightly incurvate but may be straight. Basal edge is usually thinned and ground. Stem sides were also usually ground. Note: The stem length often exceeds the length of the blade.

Broad, shallow, random flaking shaped the blade and hafting area, followed by shorter retouch along all edges. The point was finished by grinding all edges of the stem.

The type was named after Jude Hollow in Madison County, Alabama, where examples were first noted. Associations seem to indicate early Archaic or transitional Paleo period origin. Heavy patination and ground base seem to confirm this.

72-A - $45.00 72-B - $10.00 72-C - $25.00 72-D - $20.00
72-E - $10.00 72-F - $20.00 72-G - $7.00

Kays

(Pictures 73-A thru 73-C)

Kays is a medium to large, straight, stemmed point with excurvate blade. Average length ranges from 2-1/2 inches to 3-1/2 inches and from 1 inch to 1-1/4 inches in width.

Kays has a biconvex cross-section. The blade is excurvate but may be nearly straight. Shoulders are generally tapered but may be horizontal or rounded on some examples. The distal end is acute. The stem is straight with a straight basal edge. Some examples may be either slightly in - or excurvate. Basal edge is usually thinned and ground, the base sides may also be ground.

Broad, shallow to deep flaking is used to shape the blade and stem or sometimes collateral flaking. A short regular retouch is used to finish edges of blade and haft. Usually made of local material and usually patinated.

The type gets its name from Kays Landing in Henry County, Tennessee where Kneberg first recognized the point. Kays is believed to be from 2,000 to 5,000 years old. A late Archaic association is indicated.

73-A - $125.00 **73-B - $75.00** **73-C - $75.00**

Kirk Corner Notched

(Pictures 74-A thru 74-G)

This type is a medium to large sized, corner notched point, usually finely worked. Length ranges from over 1-1/2 inches to over 4 inches long and from 1 inch to nearly 2 inches in width at shoulders.

This type will usually be flattened but may be biconvex. The blade is generally excurvate but may be straight or recurvate. Edges of the blade are usually serrated and beveled on both sides of each face. Shoulders are very strongly barbed. Removal of notches at each corner leaves an expanded stem. Stem sides are straight or slightly incurvate with a basal edge either incurvate, straight or rarely excurvate. Basal edge is thinned and often ground.

Blade and stem were shaped by broad, shallow, random flaking. The edges are retouched by fine, regular pressure flaking resulting in fine serrations and a short bevel along blade edges. Notches were placed diagonally at each corner by removing a long broad flake from each face on both sides, with retouch in the notches common. The basal edge was thinned by removal of broad, shallow flakes from the base, then the edge was finished with a fine retouch.

The type is named for examples found in the North Carolina Piedmont by Coe in 1956. Estimates as much as 8,000 years B.P. exist. The type is almost always found on sites associated with early Archaic assemblages.

74-D - $300.00 **74-F - $125.00** **74-G - $450.00**

(74-G stolen, if seen, contact author!)

74-A - $950.00 74-B - $1,000.00

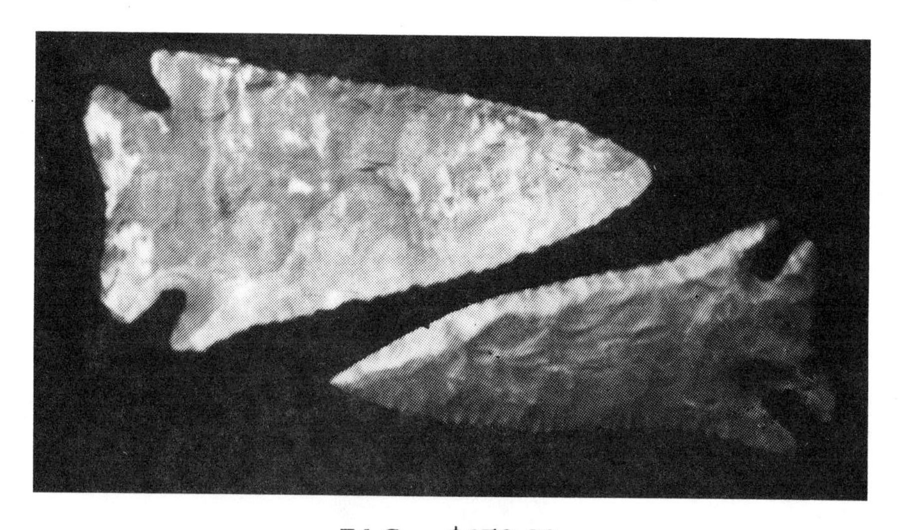

74-C - $450.00
74-E - $500.00

Kirk Serrated

(Pictures 75-A thru 75-F)

This is a medium to large, heavily serrated, stemmed point. Length ranges from about 1-1/2 inches to over 4 inches long. Examples approaching or exceeding 3-1/2 inches are all rare, but do occur. Width averages from 1 inch to 1-1/2 inches at the shoulders.

Cross-section is most often biconvex but may be plano-convex. Blade edges are often one edge excurvate with the other edge recurvate or rarely straight. Shoulders are usually horizontal or inversely tapered. Shoulders or barbs may expand. Blade edges must be serrated, with serration becoming deeper and more pronounced as the haft area becomes closer. Stem has straight sides and incurvate or straight basal edge. The base is thinned and, in rare cases, has beveled edges.

Blade and stem were shaped by broad shallow random flaking. This was followed by deep short flaking to serrate the blade edges. The stem is finished with short shallow retouch.

The type was named for sites in the North Carolina Piedmont and was first recognized by Coe in 1959. This point is believed to have occurred around 7,000 years B.P. A little later than Kirk Corner Notched but still early Archaic.

75-D - $275.00 **75-E - $125.00** **75-F - $75.00**

75-A - $2,500.00 **75-B - $125.00** **75-C - $150.00**

Knight Island

(Pictures 76-A thru 76-E)

Knight Island is a thin medium, sized point with side notches. Lengths range from 1-1/4 inches to 2-1/2 inches long and 3/4 inch to 1/2 inch in width.

The cross-section is flattened on nearly all examples or rarely plano-convex. The blade is usually excurvate but may be parallel angular.

(continued on next page)

The distal end is acute. The shoulder is straight on examples with narrow notches, or inversely tapered on examples with broader notches. The hafting area is side notched about 1/8 inch above basal edge. Basal edge is usually straight and thinned and may also be lightly ground.

The blade and hafting area are shaped by removal of broad, shallow, random flakes. All edges are then finished with fine retouch. The notches are formed by removal of broad shallow flakes from each side of haft. Notches are then finely retouched.

The type was named after Knight Island in the Tennessee River in North Alabama where the type was first recognized by Hulse. A likely relation to Jacks Reef Corner Notched exists as they are often found on same late Woodland sites and share much similarity in characteristics. A possible late Woodland date of around 1,000 B.P. is indicated.

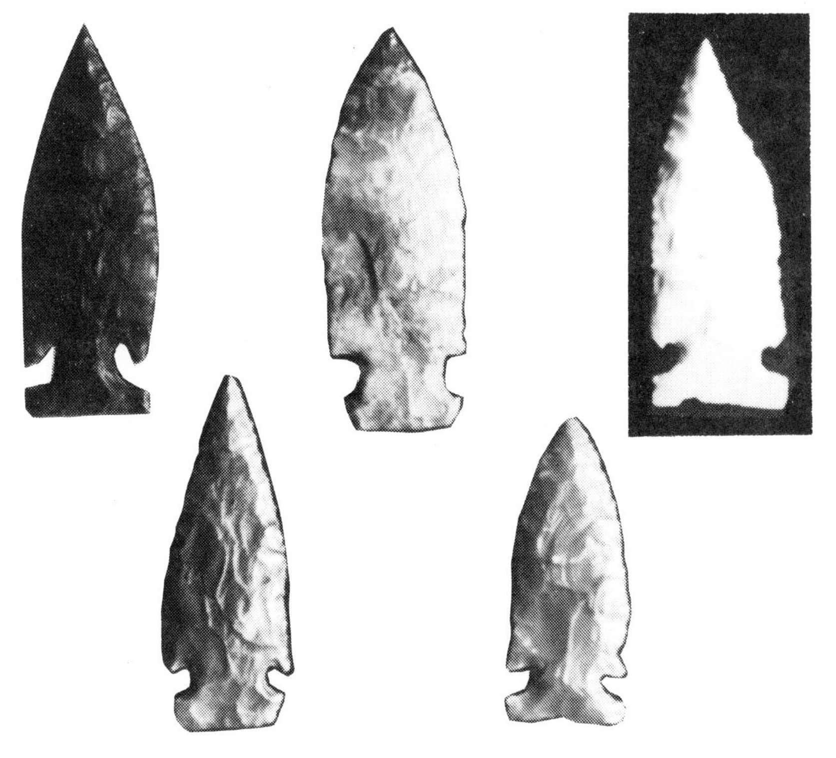

76-A - $500.00 **76-B - $175.00** **76-C - $35.00**
76-D - $50.00 **76-E - $25.00**

Lafayette

(Picture 77-A)

The blade tends to be more triangular than the Clay; notches are often not as deep or narrow as the Clay.

Tends to resemble the Kirk Corner Notched points of north Alabama and Tennessee.

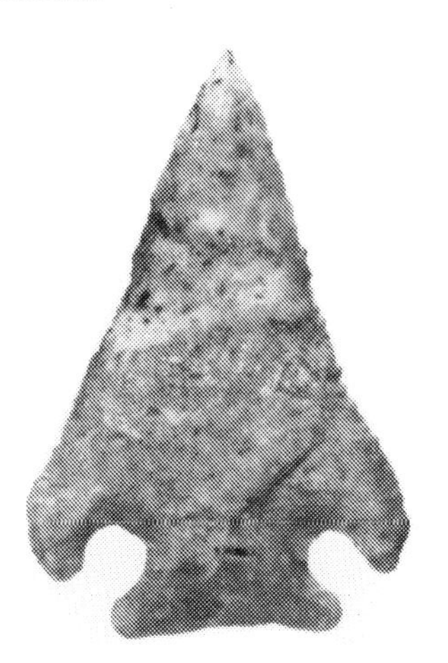

77-A - $200.00

Lauderdale

(Pictures 78-A thru 78-F)

A medium size stemmed point. Blade edges are usually recurvate, sometimes excurvate. Chipping is random. The blade edges are finished by good retouching which sometimes forms strong serrations. Shoulders are straight to barbed. The base is expanded with an oval basal edge and has the appearance of broad corner notching. The point is found on late Archaic and early Woodland sites. This point is sometimes confused with the Plevna.

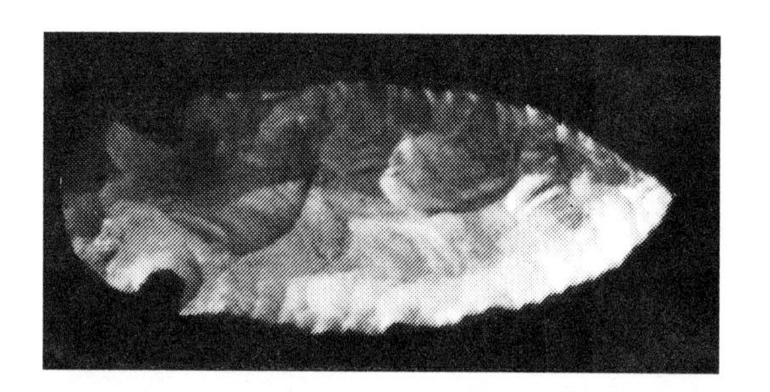

78-F - $35.00

78-C - $750.00 78-D - $45.00

78-E - $10.00

78-A - $450.00

78-B - $250.00.00

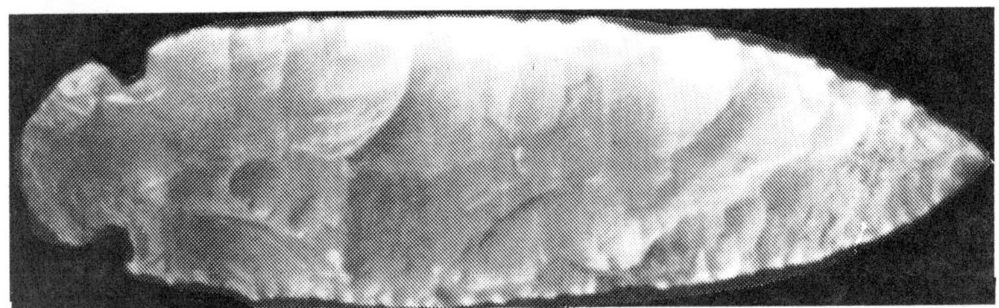

LeCroy

(Pictures 79-A thru 79-H)

The Lecroy is a medium-sized, usually serrated, bifurcated, stemmed point. Lengths range from 1-1/4 inches to about 2-1/4 inches long and 3/4 inch to 1-1/4 inches in width at the shoulders.

The cross-section is generally biconvex, but plano-convex or hexagonal may be seen in rare cases. The blade is usually straight and only rarely incurvate. Blade edges are often deeply serrated and sometimes beveled on each side of both faces. The distal end is acute. The shoulder is usually expanded on serrated examples and may be horizontal or tapered. The hafting area is usually expanded stemmed and is always deeply bifurcated. The base has expanded-rounded auricles and is nearly always ground.

The shaping of the blade usually employed removal of broad, shallow random flakes. In some cases the flaking used to form the deep serrations on the blade edges, approached collateral in appearance. The larger serrations were formed by removal of several deep flakes. The shoulders were formed by removing broad deep flakes from the side of the hafting area. The basal edge was deeply notched in the same manner. All edges of the hafting area was then finely retouched, then ground.

The type was named after the LeCroy site on the Tennessee River in Hamilton County, Tennessee and was described by Bell in 1960. LeCroy is usually found on sites with early Archaic and transitional Paleo artifacts. Distribution appears to be greatest along the Wheeler Basin of the Tennessee River in North Alabama. Indications are for an early Archaic association of around 5,000 years B.P.

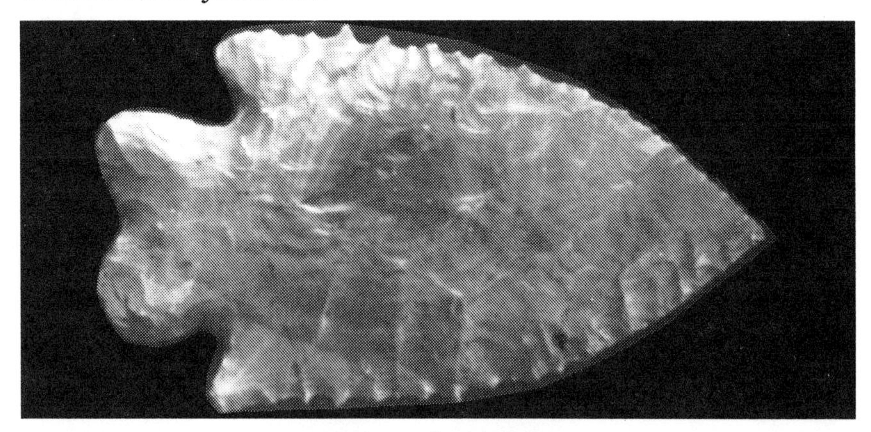

79-A - $1,500.00

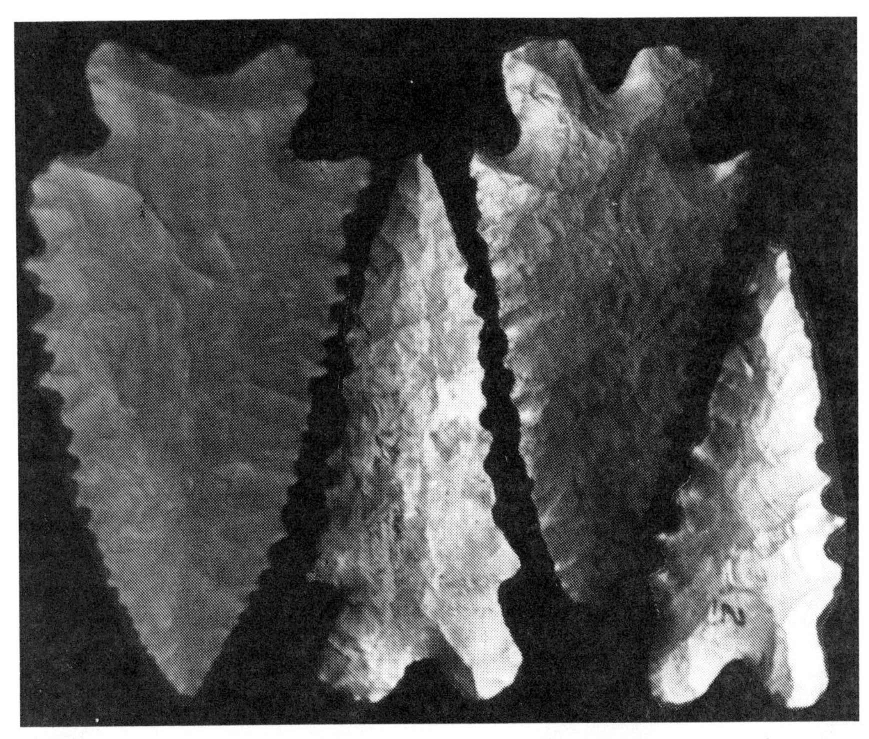

79-B- $500.00 79-C- $125.00 79-D - $150.00 79-E- $75.00

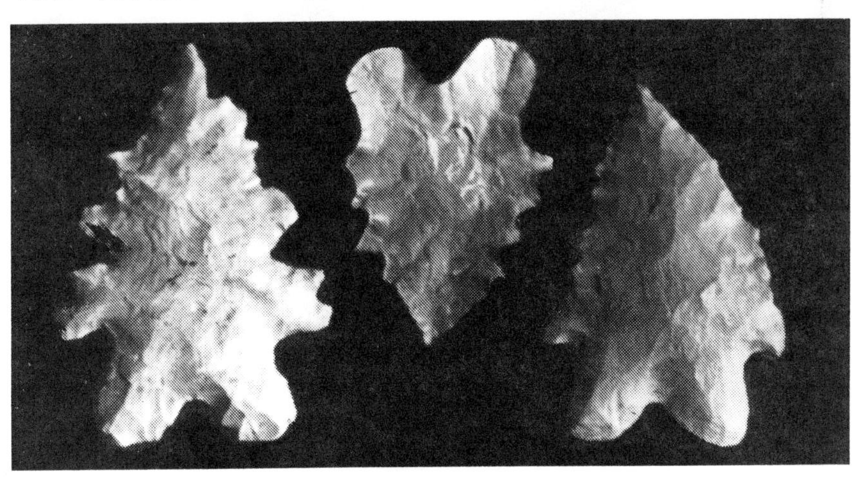

79-F $300.00 79-G $75.00 79-H $45.00

Ledbetter

(Pictures 80-A thru 80-F)

Ledbetter is a large, asymetrical, stemmed point. Average length is around 2 inches to 3 inches long and from 1-1/4 inches to 1-1/2 inches wide at shoulders.

Cross-section is biconvex. Blade edges are recurvate with opposite curvature on each blade resulting in an asymetrical blade shape. Shoulders are also asymetrical as one will be wider than the other, and may be either tapered or straight. The distal end is acute. The stem edges may be straight or slightly incurvate with a straight or rarely excurvate basal edge. Basal edge is thinned.

Broad, shallow, random flaking was used to shape the blade and hafting area. Examples may have fine secondary flaking on most of blade edges or may be completely unretouched.

The type was named after the Ledbetter site in Benton County, Tennessee. The point was recognized by Kneberg in 1956. Examples are found on sites with Archaic to early Woodland material. An Archaic period date from around 4,000 years B.P. to as late as 2,000 B.P.

80-D - $450.00 80-E - $150.00 80-F - $80.00

80-A - $450.00 80-B - $500.00 80-C - $75.00

Leighton

(Pictures 81-A thru 81-H)

A medium size, double side notched point.

Blade edges are straight to excurvate and are often serrated. The basal edge is incurvate to bifurcated.

Flaking techniques are similar to Pine Tree or Big Sandy. The Leighton point is found on early Archaic sites along with these two types.

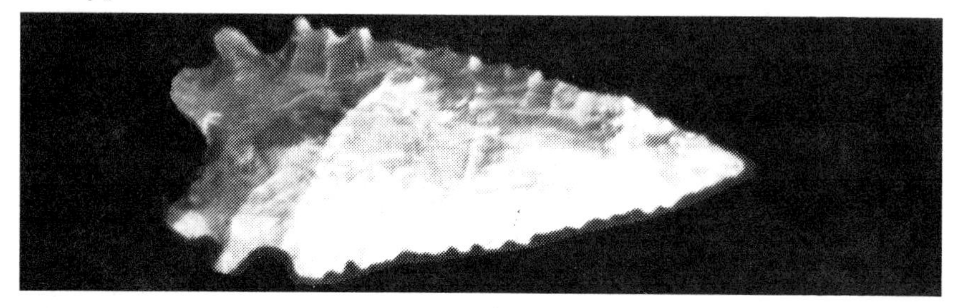

81-A - $350.

81-B - $150. **81-C - $45.** **81-D - $275.**
81-E - $25. **81-F - $75.** **81-G - $75.** **81-H - $10.**

Leighton Straight Base

(Pictures 82-A thru 82-G)

A medium sized, double notched point. Blade edges are usually straight. This type is seldom serrated but has blade edges formed by good retouching. Basal edge is straight - sometimes excurvate. One example has expanded basal barbs. Age would appear to be Archaic, possibly into the early Woodland period. A few examples have pointed stems.

82-A - $100.00

82-B - $45.00 **82-C - $200.00** **82-D - $25.00**

82-E - $100 **82-F - $35.00** **82-G - $50.00**

Lerma

(Pictures 83-A thru 83-E)

Lerma pointed base is a medium to large sized point with pointed base and excurvate blade. Average length will range between 2 inches and 4 inches long and around 2 inches to 1-1/2 inches in width. Cross-section may be either biconvex or plano-convex. From the distal end to the widest point is usually excurvate but may be straight.

The distal end is acute. The haft area is either broad or acute pointed.

Shaping employs removal of shallow well controlled flakes. Secondary flaking was used to finish most examples. Haft may be undiscernible from rest of point.

Lerma was first recognized at the Canyon Diablo site of Tamaulipas, Mexico. The type was found in sites with mammoth remains on the site near Mexico City. In north Alabama the type is found in association with Big Sandy I, Dalton, Crawford Creek and Lerma Round Base. The Round Base variety appears to be slightly younger. From these associated types it would appear that is is an early Archaic to transitional point type.

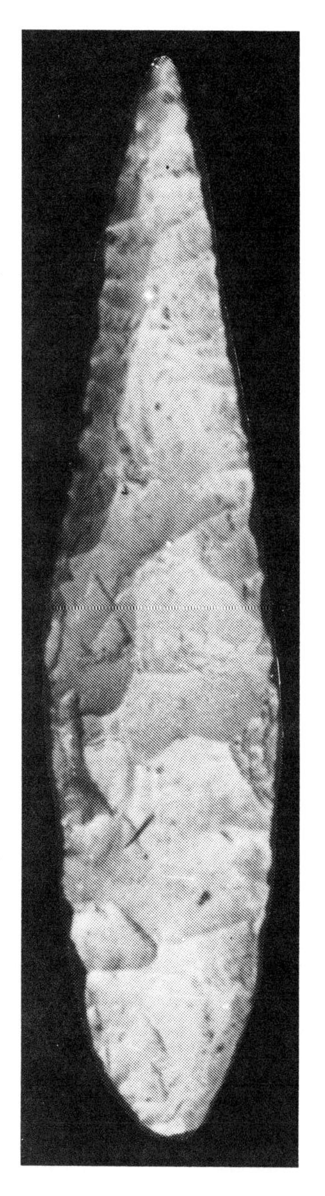

83-A
$250.00

83-B - $125.00 83-C - $200.00 83-D - $100.00 *Burrin Tip*

83-E
$50.00

Levy

(Picture 84-A)

A medium sized point. Usually thicker than the Newnan types. Stem juncture is rounded as in the Pickwick points of the Tennessee River Valley. Flaking is random. Base is straight to incurvate. Shoulders are often barbed.

81-A - $300.00

Limestone

(Pictures 85-A thru 85-D)

Limestone is a small to medium sized stemmed point with an incurvate basal edge. Length averages from 1-1/2 inch to 2 inches long and 1 inch to 1-1/2 inches in width.

Cross-section is biconvex. The blade is straight with an acute distal end. The shoulders are usually tapered but, rarely, are horizontal or rounded. The stem ranges from straight to somewhat expanded. Sides of stem are either straight or incurvate. Basal edge will always be incurvate and thinned.

(continued on next page)

The shaping of the blade and stem employs broad, shallow, random flaking. Blade edges were usually finished with fine to rather crude secondary flaking. The stem is shaped by the removal of large deep flakes from the corners of the blade. This area was then given a fine retouch. The basal edge was thinned and then often retouched. Many examples are of Bangor flint.

The type was named by Cambron for examples found in a shell mound in Limestone County, Alabama. There appears to be a late Archaic or early Woodland age for Limestone points.

85-A-$125.00 **85-B-$75.00** **85-C-$45.00**
85-D-$15.00

Little Bear Creek

(Pictures 86-A thru 86-E)

This is a medium to large-sized, long stemmed point. Average lengths range from 2 inches to about 4 inches long and from 3/4 inch to just over 1 inch in width.

Cross-section is biconvex. Blade may be excurvate or occasionally straight. The distal end is acute. Shoulders will either be tapered or horizontal. The stem is either straight or contracted with straight sides. Basal edge is straight and may not be finished. Stem edges are usually ground.

The blade and hafting area is shaped by deep random percussion flaking. There may also be some secondary percussion flaking used to finish edges.

The type was named after Little Bear Creek, site Ct 8, where it was first recognized as a type. Appearance seems to have taken place during shell mound Archaic period and reached its apex during early Woodland. Earliest probable date was around 5,000 years B.P.

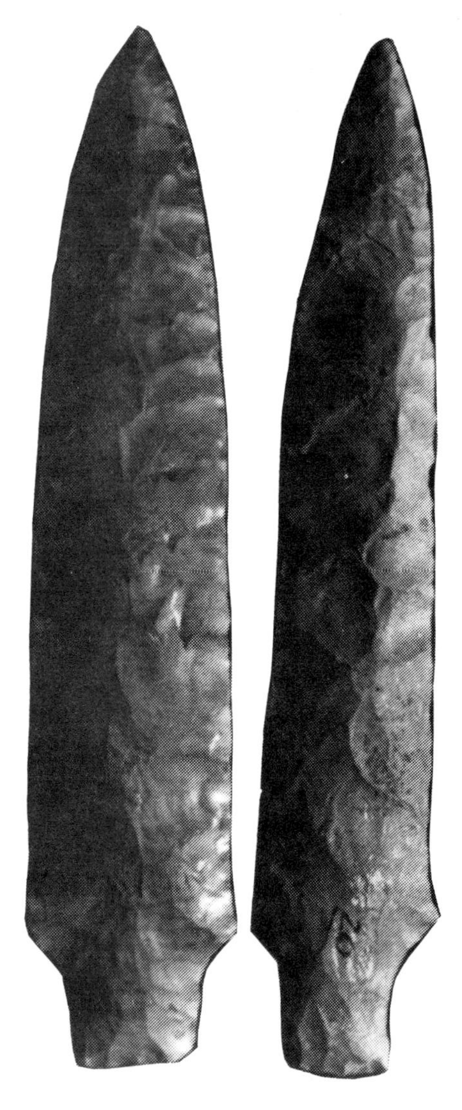

86-A - $500.00 **86-B-$450.00**

86-C - $250.00 **86-D - $70.00** **86-E - $45.00**

Lost Lake

(Pictures 87-A thru 87-L)

Lost Lake is a medium to large-sized, corner notched point that is beveled one side of each face. Average length ranges from 1-3/4 inches to over 3-1/4 inches long. Points over 3-1/2 inches are encountered but are quite rare. The width may range from 1-1/4 inches to nearly 2 inches.

Cross-section is always rhomboid. Blade edges will usually be straight or occasionally excurvate or recurvate. Blade edges may be serrated on some examples but will always be beveled on one edge of each face. The shoulder barbs are usually long and simple, but may be rounded, acute or expanded. Distal end is usually acute but may be broad.

(continued on next page)

The hafting area has an expanded stem that is the result of notches placed diagonally at the corners or rarely, diagonally notched from base. The stem sides are straight or incurvate. The basal edge of the stem may be incurvate, straight or slightly excurvate. Basal edge is thinned and usually ground with the exception of some straight-based examples.

The blade faces were shaped and flattened by removal of broad, shallow, and random flakes. The steep, regular flakes used to bevel the blade edges were removed from one side of each face. This sometimes left a finely serrated blade edge. The notches were started by removal of small flakes for about 1/3 of the depth of the finished notch, the notches were then terminated by removal of a broad flake in each notch on each face. Some retouch is applied in the notches. Shallow, broad flakes were removed to thin the basal edge.

The type was named for the Lost Lake area located in Limestone County, Alabama, where many examples were found. Examples found in north Alabama come from pre-shell mound sites, suggesting an early Archaic association before 5,000 years B.P.

Editors Note: Originally all points that were corner notched were identified as Cypress Creek points by Kneberg and Lewis. In an article in the First Ten Years of Alabama Archaeology (page 457), Cambron suggests that only the point now known as Lost Lake be identified as Cypress Creek. Later the name was dropped when the Lost Lake was reidentified.

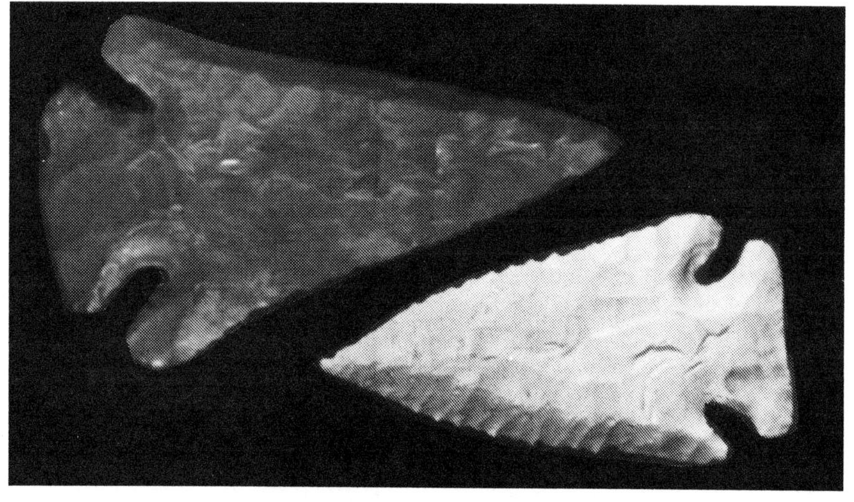

87-A - $450.00 **87-B - $400.00**

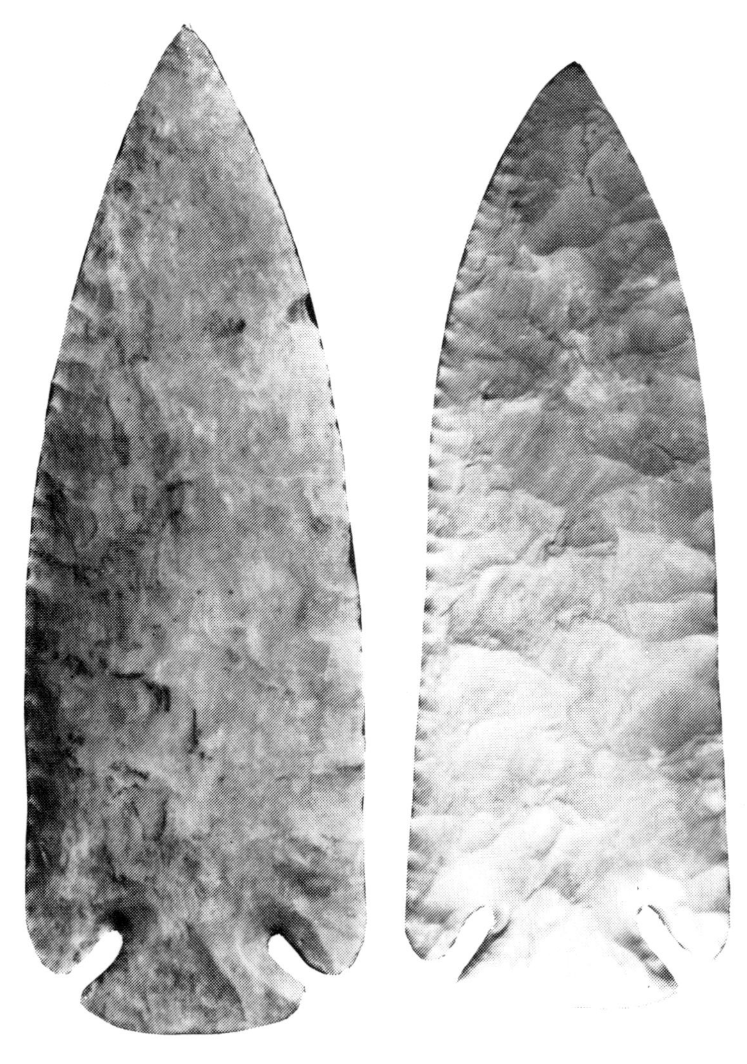

87-C - $2,500.00

87-D - $2,200.00

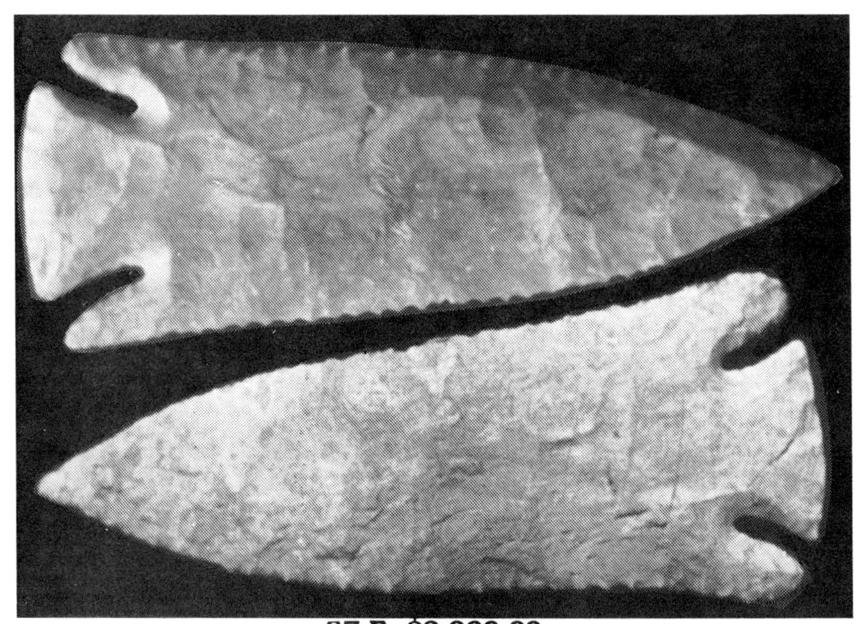

87-E -$2,000.00
87-F -$1,800.00

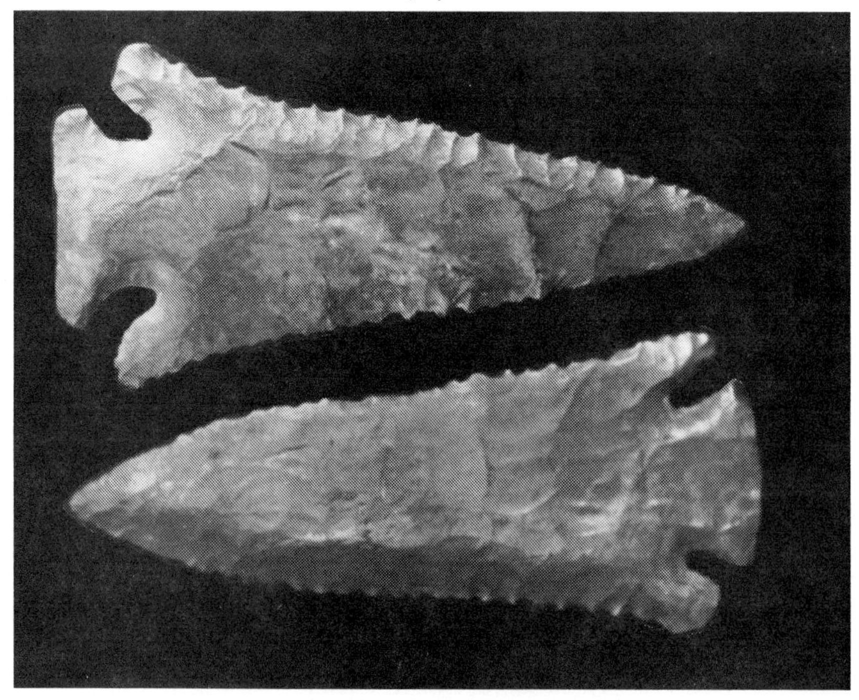

87-G-$1,500.00
87-H -$500.00

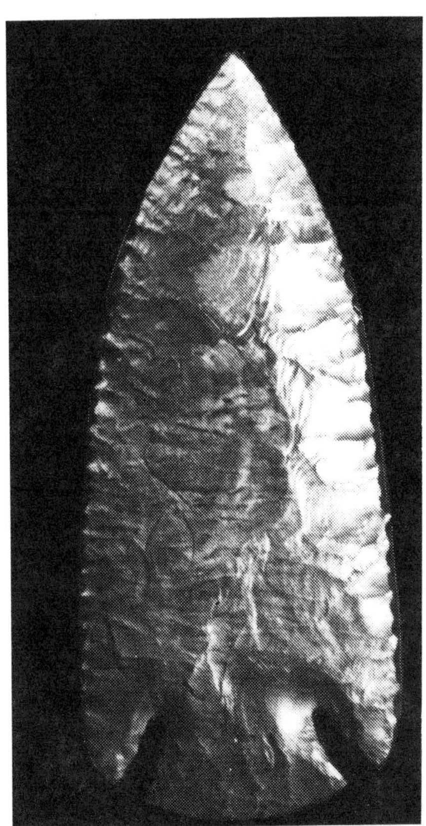

87-E-$2,000.00

87-F-$1,500.00

87-I-$750.00 87-J-$1,000.00

Madison

(Pictures 88-A thru 88-H)

Madison is a small, thin, triangular point. Lengths average from about 1/2 inch to around 1-1/2 inches long, widths range from 1/2 inch to 3/4 inch.

The cross-section is flattened and thin. Blade edges are nearly always straight but may be slightly excurvate. Hafting area is not defined from the blade, though some portion of the basal area of the blade was used in hafting. Basal edge is usually straight, but may be slightly incurvate. Base is always thinned.

The blade was shaped by removal of broad, shallow, random flakes. Shaping of the blade and base was finished with fine secondary flaking. The type was first referred to as Mississippi Triangle, by Scully, who later, changed it to Madison. The type appears to be widely distributed throughout the Eastern U.S. The type has been linked to prehistoric Iroquois in New York and in Alabama with prehistoric Creek. Associations in Alabama seems to suggest a late Woodland origin with most occurring during the Mississippian period.

88-A-$15.00 **88-B-$50.00** **88-C-$35.00**
88-D-$100.00 **88-E-$35.00** **88-F-$25.00** **88-G-$35.00**
88-H-$10.00

Maples

(Pictures 89-A thru 89-D)

Maples is a large, broad stemmed point with incurvate base. Average length is from 2-1/2 inches to over 4-1/2 inches long and from 1-1/4 inches to 2-1/2 inches in width.

The cross-section is biconvex. The blade edges most often are excurvate but may be straight. Distal end is usually acute. Shoulders are usually tapered. The short broad stem is usually contracted, rounded with excurvate sides. Basal edge may either be excurvate or straight, often thinned and occasionally ground.

The blade and stem were shaped by broad, shallow to deep flaking. Only sparse retouch was used, sometimes appearing on only one edge of each face in some points. Other examples will show extensive retouch along blade edges. Basal edge and stem edges usually show only light retouching. Examples are made from local material with Ft. Payne chert preferred.

The type was named for sites near Maples Bridge on the Elk River in Limestone County, Alabama. Maples are found in association with middle and late Archaic points and tools. A date prior to 4,000 years B.P. is suggested.

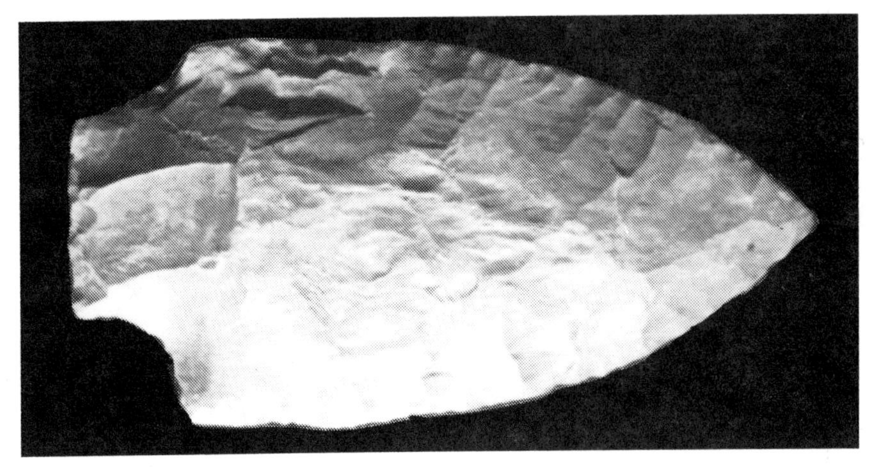

89-C - $75.00

89-A - $800.00 89-B - $500.00

89-D - $25.00

Marion

(Pictures 90-A thru 90-B)

Part of the Newnan cluster. Similar to the Newnan except the stem is rounded. Similar to an Adena Narrow Stem except the stem is shorter than that of an Adena.

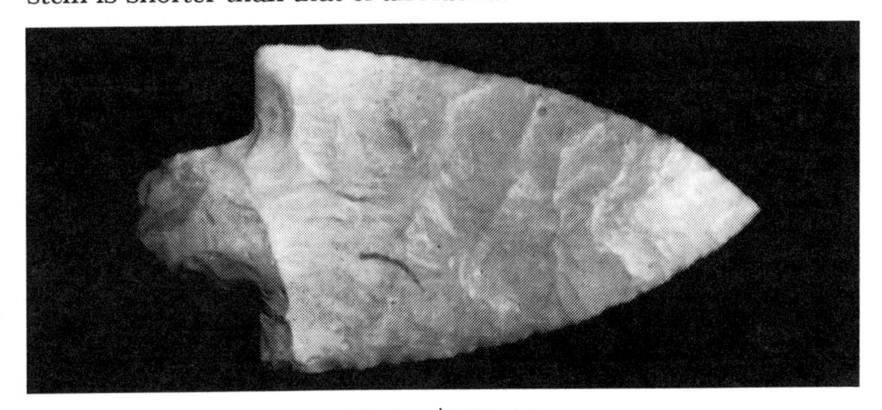

90-A - $350.00

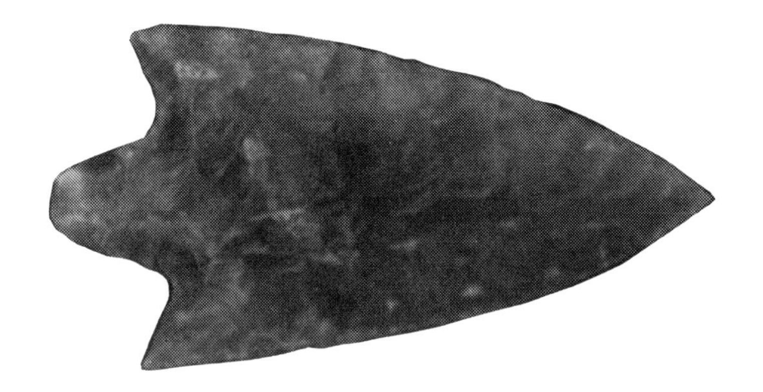

90-B - $750.00

McIntire

(Pictures 91-A thru 91-D)

McIntire is a medium-sized, expanded-stem point with excurvate blade edge. Average length is from 2 inches to about 2-1/2 inches long and between 1-1/4 inches and 1-1/2 inches in width.

Cross-section is biconvex. Blade edges are excurvate with some examples having one straight blade edge. Shoulders are often horizontal but may be tapered or inversely tapered with slight barbs. The distal end is acute. The stem is expanded with incurvate side edges. The base of the stem is usually straight and thinned but rarely may be slightly excurvate.

Broad, shallow, random flakes were removed to shape the blade and stem. Retouch along blade edges is short and sometimes rather deep. The stem was formed by the removal of broad deep flakes from corners of the blade. All edges of the stem were retouched. Local nodular flint sources were used as demonstrated by rind patina found on basal edges of many examples.

The type is named for sites around McIntire ditch on the north bank of the Tennessee River near Decatur, Alabama. The type was first noted by Hulse. McIntire is found associated with Archaic shell-mound cultures. The Limestone point is often found on sites with this type, and shares several similarities in form and flaking. One area that separates the two types is the basal edge of the stem. Limestone is always quite incurvate whereas the McIntire basal edge will be straight or nearly straight. Surface finds suggest a middle to late Archaic cultural tie.

91-A - $150.00 **91-B - $75.00** **91-C - $75.00** **91-D - $25.00**

McKean

(Picture 92-A)

McKean is a small to medium, excurvate bladed point with incurvate, thinned base. Average length is from 1-1/4 inches to 2-1/2 inches long and 1/2 inch to 3/4 inch in width.

The cross-section is usually biconvex, but in some examples may be flattened. The blade is usually excurvate but may be nearly straight or recurvate. The auriculate base is usually contracted with rounded auricles or may, in some cases, be parallel or expanded rounded. Basal edge is incurvate and well thinned. Distal end is acute.

Random flaking was employed to shape the blade and hafting area. In a small number of examples, flaking may be collateral. The blade edges are usually retouched. Examples from the northwestern U.S., are transverse-obliquely flaked.

(continued on next page)

The type was named for examples found in the area of Keyhole Reservior in northeastern Wyoming, by Wheeler in 1952. There is some indication the McKean is a derivative of the Wheeler points. Estimated antiquity probably should begin about 9,000 years B.P. to possibly 4,000 years B.P. in the southeastern U.S.

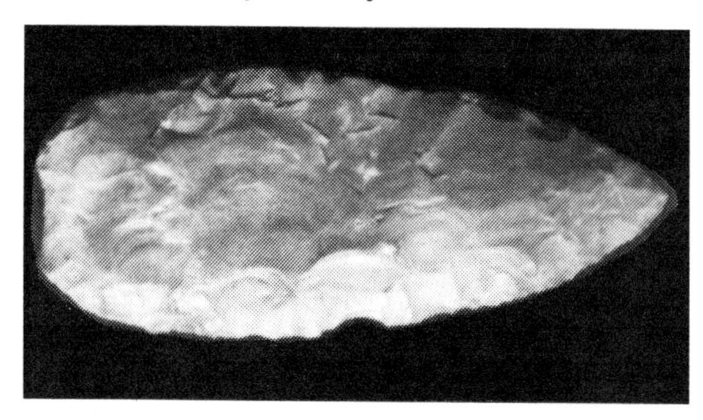

92-A
$35.00

Morrow Mountain

(Pictures 93-A thru 93-I)

The Morrow Mountain is a medium-sized point with excurvate blade and rounded stem. Average lengths are from 1-1/4 inches to a little over 2-1/2 inches long and from 3/4 inch to just over 1 inch in width.

Cross-section is biconvex. Blade edges are excurvate but will be recurvate on examples that are expanded at the shoulders. Shoulders are narrow, and are usually horizontal but in some examples may be inversely tapered and occasionally expanded. The distal end can be acute, micronate, or, in rare instances, broad. The hafting area is a rounded, or rarely, pointed stem. On some examples, stem edges may be ground. Some examples may be serrated on blade edges.

The blade and hafting area was shaped by removal of narrow, shallow, random flakes. Fine retouch was used on all edges to finish shaping. On about half of examples, short, fairly deep, flakes removed from all or part of the blade edge resulted in fine serrations. Local flints and cherts were used.

The type was named for Morrow Mountain in the North Carolina Piedmont, by Coe in 1959 as suggested by James B. Griffin. Examples recovered in context as well as associations encounters in surface collections indicate an early Archaic emergence. Sometimes in excess of 6,500 years old. Variants are Straight Base and Round Base. Also thought to be associated with Bascum Blade.

93-A -$45.00 93-B -$45.00 93-C -$35.00

93-D - $25.00 93-E - $55.00 93-F - $35.00

A-D - Straight Base
E-F - Morrow Mountain

93-G - $20.00 **93-H - $25.00** **93-I - $10.00**

G - Morrow Mountain; H-I - Round Base

Motley

(Pictures 94-A thru 94-E)

Motley is a medium-sized, expanded stem point with large side or corner notches. Average length of Motley is between 2 inches and 2-3/4 inches long and between 1 inch and 1-1/4 inches wide.

Cross-section is either flattened or biconvex. Blade edges are usually straight or only slightly excurvate. Some examples are noted to have one blade straight with the other slightly excurvate. Shoulders range from inversely tapered or tapered to horizontal, depending on the slope of the notches. The distal end is acute. Deep side notches or deep corner notches leave an expanded stem for hafting. The sides of the stem are incurvate with a straight basal edge. Basal edge is usually thinned but only rarely ground.

Well controlled, shallow to deep, random flaking was used to shape the blade and stem faces. Blade edges nearly always show secondary flaking along edge of blade, usually short fairly deep flakes. Notches were formed by removal of large deep flakes followed in most cases by secondary flaking to complete shaping of the notches. It should be noted some points are side notched, some are more corner notched, and still others have one of each type notch. Local materials were utilized.

(continued on next page)

The type was named for Motley Place in the northeastern corner of Louisiana. Radio-carbon samples from the poverty point culture sites of the lower Mississippi Valley yielded dates from about 1,300 B.C. until about 200 B.C. This cultural focus seems to be coincidental with Motley as evidenced by its heaviest distribution being in this area. This point type appears on predominately Archaic sites of Alabama, Illinois, Kentucky and Tennessee. Evidence from sites in Alabama indicates a beginning in late Archaic, increasing occurrence into the early Woodland period.

94-A - $200.00 **94-B - $400.00** **94-C - $125.00**
94-D - $50.00 **94-E - $50.00**

Mountain Fork

(Pictures 95-A thru 95-C)

Mountain Fork is a small, narrow point, often somewhat crude with a thick stem. Averaging between little over 1 inch to about 2 inches in length and from 3/4 inch to 1/2 inch in width at the shoulders.

Cross-section may be biconvex or, less often, slightly median ridged. Blade edges may be either straight or excurvate with an acute point. Shoulders are tapered and narrow. The stem may be straight or tapered. The basal edge is either straight or slightly excurvate. The basal edge is usually left unfinished but may be somewhat thinned. Local materials such as Bangor nodules were used.

Faces of the blade and stem were shaped by removal of short, deep, random flakes. Retouch on the blade edges was an even, short, deep flaking. The quality of available working material may account for some examples being somewhat crude.

This point type derives its name from points found on sites along Mountain Fork Creek in Madison County, Alabama, where the type was first recognized and named by Cambron. Surface sites tend to yield such types as Flint River Spike, Bradly Spike, and Swan Lake in conjunction with this type. From best indications, they occur during middle to late Woodland period. This type is often found with impact fractures on the distal ends, causing some to consider them possibly arrow points.

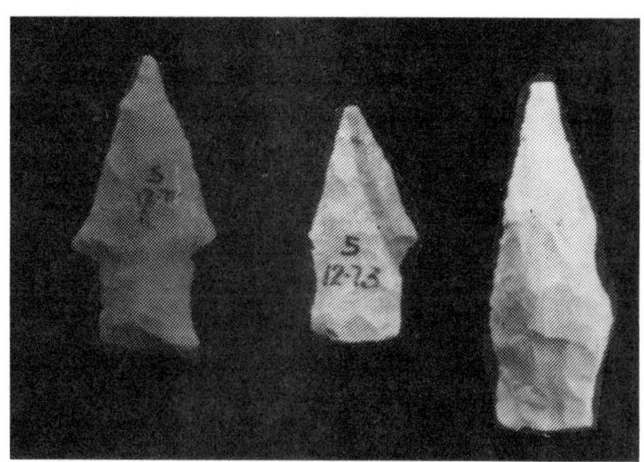

95-A - $10.00

95-B - $3.00

95-C - $2.00

Mud Creek

(Pictures 96-A thru 96-E)

Mud Creek is a medium-sized point, having an expanded stem and acuminate distal end. Average length is from 1-1/2 inches to slightly over 2-1/2 inches long and from 7/8 inch to 1-1/8 inches wide.

Cross-section is biconvex. Blade edges are always excurvate with an acuminate or extremely acute distal end. Shoulders are generally tapered but in some examples may be horizontal or sometimes rounded. The stem is expanded, usually with straight side edges. The basal edge may be straight or excurvate, and is usually thinned. A large portion have some grinding on the edges of the hafting area.

The blade and stem area is formed by removal of broad, shallow, random flakes. Blade and stem edges were retouched with small, rather deep flaking. Though most stems are thinned, some may be only crudely finished, sometimes retaining some nodule rind.

This type is named for sites in the vicinity of Mud Creek in Limestone County, Alabama, where the type was first noted. Evidence gained from excavation and surface collections from north Alabama seem to indicate a strong point type from the late Archaic period into early Woodland.

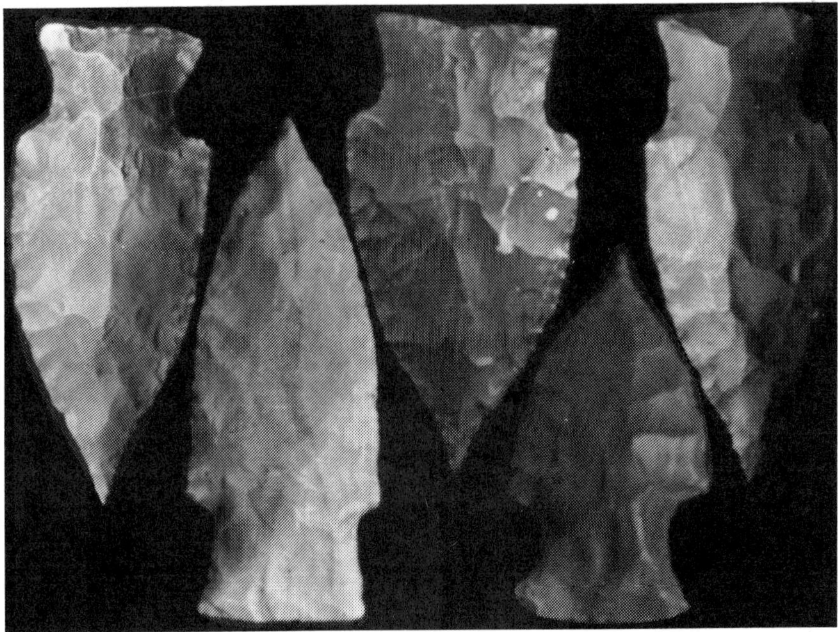

96-A - $150.00 **96-B - $100.00** **96-C - $50.00**
96-D - $45.00 **96-E - $65.00**

Mulberry Creek

(Pictures 97-A thru 97-F)

Mulberry Creek is a medium to large sized point, with excurvate blade and stemmed base. Length averages from 3 inches to over 4-1/2 inches long and 1 inch to 1-1/2 inches at the widest point.

Cross-section is biconvex. Blade edges are always excurvate with the widest point near the middle of the blade. Blade edges are sometime finely serrated. Shoulders are fairly narrow and usually tapered. Shoulders on some examples are asymetrical. The stem tends to be tapered or straight, but may rarely be expanded. Basal edge is generally excurvate and poorly thinned, but also may occasionally be straight. Stem edges are usually well ground.

Broad, shallow, well controlled random flaking shaped the blade and stem faces. Blade edges usually show fine retouch. Stem edges may be retouched with flaking a bit cruder than on the blade edges.

The type is named for sites near Mulberry Creek in Colbert County, Alabama Valley in north Alabama. Some smaller numbers have been reported in Missouri and Illinois. Indications of an emergence in Middle Archaic, climaxing in late shell-mound Archaic, with an end during early Woodland, is suggested by evidence from sites in north Alabama.

97-A- $50.00

97-B - $300.00
97-C- $75.00
97-D- $250.00 97-E - $200.00 97-F - $125.00

Nebo Hill

(Pictures 98-A thru 98-B)

A medium to large lanceolate point type. Sides are more parallel than the Sedalia. Basal edge may be straight or slightly concave or convex. Chipping is often collateral but may be transverse or even random. Cross-section is medium ridged. Age: early Archaic.

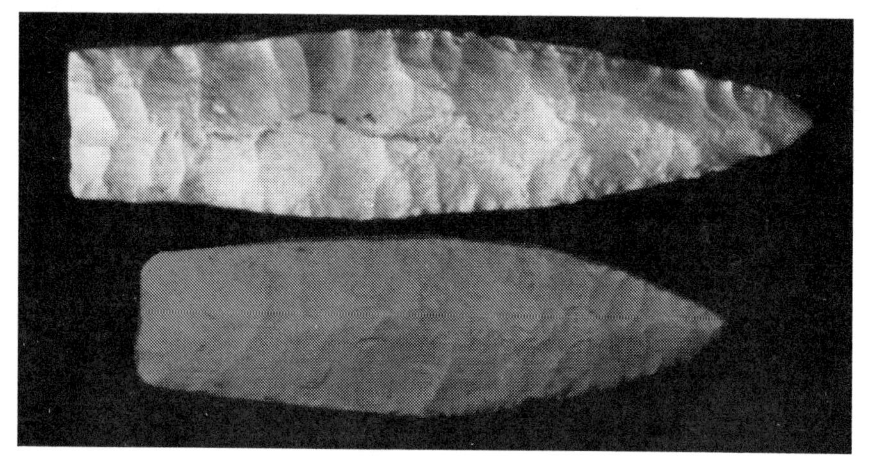

98-A - $175.00
98-B - $95.00

New Market

(Pictures 99-A thru 99-C)

New market is a medium sized, round, stemmed point with narrow blade and expanded shoulders. Average length is from 1-3/4 inches to 2-1/4 inches long and from 1/2 inch to 3/4 inch in width.

The type has a biconvex cross-section. Blade edges are usually straight but may be excurvate. Distal end is sharply acute. Shoulders are usually expanded and narrow but may be tapered. The hafting area is a contracted-rounded stem with thinned edges. Stem sides tend to be straight and basal edge is always excurvate.

Blade and stem faces are shaped by fairly deep, random flaking with deep retouching on all edges. The expanded shoulders are left from original blade as a result of this deep retouch, skipping the small place between blade and stem. Expanded shoulders are absent in examples without thinning retouch on the stem.

(continued on next page)

The type is named for New Market site near New Market, Alabama, where examples were first noted. This type was in the past classified as a Randolph point. After closely comparing specimens from north Alabama with the historic period, Randolph point, described by Coe from North Carolina, it became obvious they weren't of the same type. Aside from physical differences, there also appears to be an earlier cultural association in the New Market type. The type is found on sites with point types such as Swan Lake, Flint River Spike and Bradly Spike and likely show a late Woodland period occurrence.

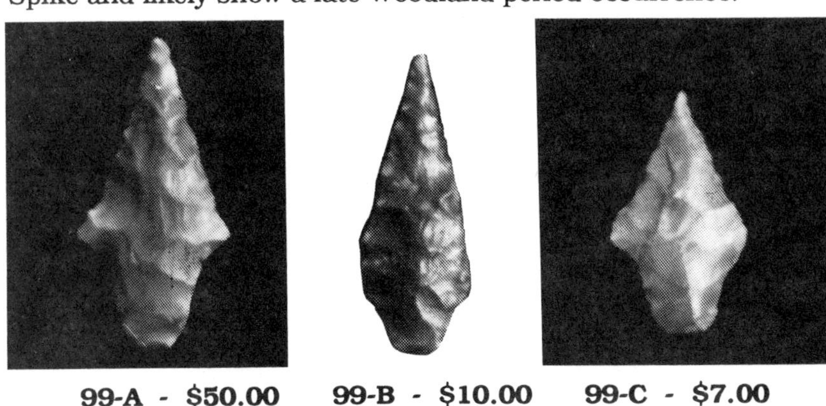

99-A - $50.00 99-B - $10.00 99-C - $7.00

Newnan

(Pictures 100-A thru 100-D)

A medium to large well made usually thin point type. The type has a constricted square-based stem. The shoulders may be straight or barbed. Blade edges are excurvate. The points of the Newnan Cluster are in the middle Archaic period.

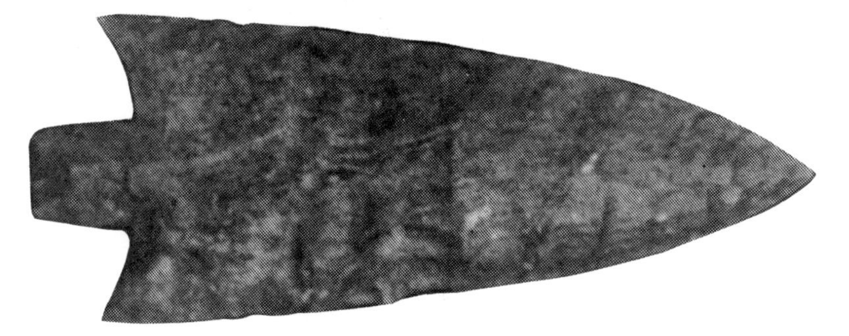

100-A - $1,500.00

100-B - $2,500.00

100-C - $1,200.00

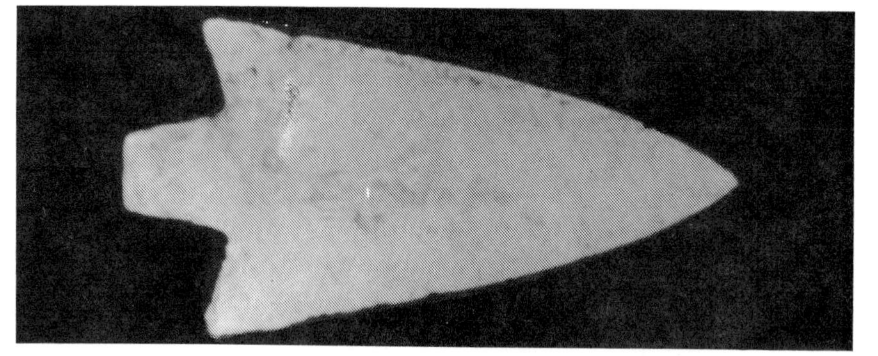

100-D - $500.00

Identification of Artifacts on Front Cover

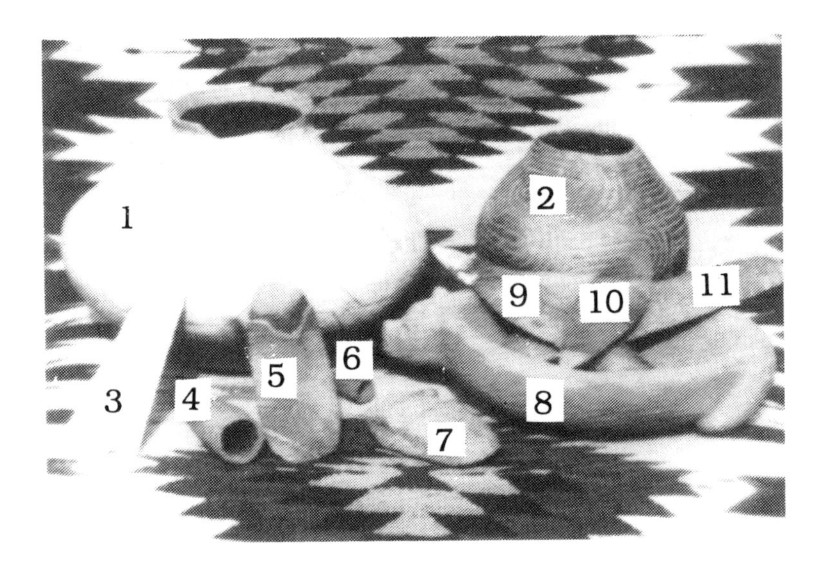

1 Southwestern Olla
2 Irene Incised Seed Jar
   Stylized Alligator Design
3 Dickson 7-1/4"
4 Pottery Tube Pipe
5 Cache Blade Made of Pickwick Tri-Color
   Jasper
6 Miniature Sandstone Tube Pipe
   Alexander Engraved
7 Southern Trophy Axe
8 Sandstone Bowl - Bear Effegy
9 Morse Knife - Red and White Jasper
10 Unnotched Turkey Tail
11 Archaic Blade

TOP ROW
1 - 2 Eccentric Points

CENTER - LEFT CORNER -CLOCKWISE
3 Spring Creek 3-1/2"
4 Mulberry Creek
5 - 8 Copena, Longest 5"
9 Copena Triangular
10 Clifton Blade
11 Pickwick
12 - 13 Wade
14 - 15 Pickwick
16 Wade
17 Smithsonian
18 Turkey Tail 3-3/4"
19 - 22 Waubesa
23 Adena Narrow Stemmed
24 - 25 Waubesa
26 Adena Narrow Stemmed
27 Waubesa
28 Adena Narrow Stemmed
29 Waubesa
30 Adena Narrow Stemmed
31 - 33 Cotaco Creek , Center 4-1/8"
34 Little Bear Creek
35 - 36 Cotaco Creek

BOTTOM
37 Flint Creek
38 Cotaco Creek
39 Crawford Creek

LOWER LEFT CORNER - END OF DESIGN
1 Madison 3/4"
2 Hamilton
3 Madison
4 Camp Creek
5 Unnamed
6 Hamilton
7 Sand Mountain
8 - 9 Hamilton
10 Madison
11 Hamilton
12 Jacks Reef Pentagonal
13 Madison
14 Hamilton
15 Jacks Reef Corner Notched
16 Bradley Spike
17 Guntersville 2-1/2"
18 - 19 Jacks Reef Pentagonal
20 - 22 Jacks Reef Corner Notched
23 Knight Island
24 Jacks Reef Corner Notched
25 Alba
26 Camp Creek
TOP POINTS IN CENTER LEFT TO RIGHT
27 Knight Island
28 Knight Island
29 Jacks Reef Corner Notched
CENTER TOP TO BOTTOM
30 Haskel 2-1/4"
31 Unnamed
32 Pottery Sun Disk 2-5/8" diameter
TOP LEFT OF CENTER DESIGN END OF CIRCLE
33 Camp Creek
34 Hamilton
35 Greenville
36 - 37 Hamilton
38 - 39 Jacks Reef Pentagonal
40 Sand Mountain
41 Madison
42 - 51 Cache of Guntersvilles
52 - 56 Cache of Knight Islands
57 Madison
58 Guntersville
59 Jacks Reef Pentagonal
60 - 62 Guntersville
63 Madison
64 Guntersville
65 Madison
66 Camp Creek
67 - 68 Madison
69 Guntersville
70 - 71 Hamilton
72 Sand Mountain
73 Hamilton
74 - 75 Jacks Reef Pentagonal
TOP RIGHT DESIGN - LEFT TO RIGHT
76 Knight Island
77 Jacks Reef Corner Notched
78 Knight Island
79 - 81 Jacks Reef Corner Notched
82 - 85 Jacks Reef Pentagonal
86 Nolichucky 2-3/8"
87 - 89 Jacks Reef Pentagonal
90 Knight Island
91 Jacks Reef Corner Notched
92 Knight Island
93 - 94 Hamilton
95 Fort Ancient
96 Camp Creek
97 Madison
98 Madison
99 Hamilton
BOTTOM ROW - LEFT TO RIGHT
100 - 101 Jacks Reef Drill
102 Hamilton 1-5/8"
103 Jacks Reef Corner Notched Drill
104 Camp Creek Drill

TOP LEFT CORNER

1 Elk River 7-3/4"

TOP RIGHT CORNER

2 Elk River 6"

CENTER ROW

3 Dickson 7-1/4"
4 Copena 8-3/8"
5 Morse Knife 8-5/8"
6 Morse Knife 9-3/8"
7 Unnotched Turkey Tail 8-3/4"
8 Benton Broad Stemmed 7-5/8"
9 Caddo Knife 7-1/2"

LOWER LEFT CORNER
LEFT TO RIGHT

10 Plevna 6-1/2"
11 Duck River Sword 8-1/2"
12 Base of Double Notched Turkey Tail
13 Benton Stemmed 6"

TOP ROW - LEFT TO CENTER

1 Lost Lake
2 Lost Lake
3 Cobbs Triangular
4 Lost Lake
5 - 20 Wheelers - Pictured in section on wheelers
21 Kirk Corner Notched
22 Kirk Stemmed Serrated
23 Pine Tree Corner Notched
24 Pine Tree Corner Notched

TOP CENTER OF OVAL - CLOCKWISE

25 Clovis 4"
26 Quad
27 Clovis
28 Greenbrier Dalton
29 Harpeth River
30 Plevna
31 Plevna
32 - 34 Big Sandy
35 Hardaway Dalton
36 - 42 Greenbrier
43 - 45 Southern Hardin
46 - 48 Decatur
49 - 51 Clovis
52 Beaver Lake

PRICING INFORMATION

The prices given in this guide are the top prices you would expect to receive for an artifact of the quality and size pictured. (In a few instances the photographs are not to actual size and the size is given in the caption.) Actual prices you may receive will depend on many factors. You must first find a buyer who is willing to pay what the seller asks. Another factor is location. An artifact will generally bring more in the area where it was found. Also an artifact brings more in an area which has a high concentration of collectors. Naturally an artifact show is a very good place to sell an artifact.

If you are selling to a dealer, you can expect to receive less than book price. Dealers usually pay 1/2 to 1/3 of book value. Some dealers give less.

Another determining factor is the eye appeal of the artifact. An extremely well made artifact which happens to be made of an ugly material will bring much less than one which is very colorful. One of the most sought after materials today is the Tri-Color Jasper of the Tennessee River Valley. Some authors refer to this as Horse Creek Chert. This is a local name given to it by collectors in Hardin County, Tennessee. The correct name given to it by the U.S. Geological Survey is Pickwick Tri-Color Jasper. Archaeologists in the Southeast refer to it as Pickwick Chert.

Along the Ohio River the material most in demand is Flint Ridge and in Florida it is Agatized Coral.

Nodena

(Pictures 101-A thru 101-G)

Nodena is a small to medium, narrow point with a rounded base. Average length of the type will be 1-1/4 inches to 2-1/4 inches long and about 5/8 inch in width.

The cross-section is biconvex. The distal end is acute. Blade edges are excurvate with little or no difference between blade and hafting area. Basal edge is usually rounded or occasionally acute.

Blade and haft area was shaped by broad, shallow, random flaking. Some points show a slight median ridge. Fine, narrow, shallow retouch was applied to the blade and basal edges.

The type is named for the Nodena site in Arkansas, were it was recognized as a type. The type has been referred to as Willowleaf type by collectors for some time. The more dense concentration of this type occurs in eastern Arkansas along the larger river systems in that area. Bell gives us an estimated antiquity around 1,400 to 1,600 A.D. or late prehistoric culture.

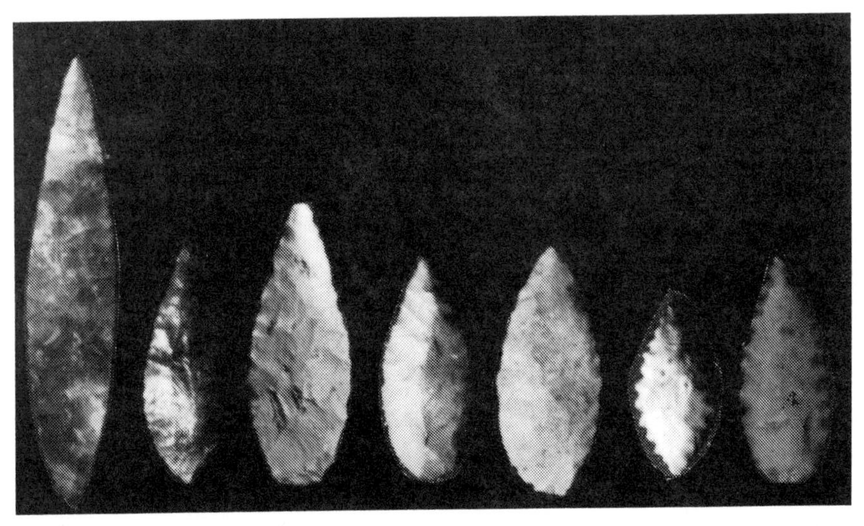

101-A	101-B	101-C	101-D	101-E	101-F	101-G
$75.00	$25.00	$5.00	$10.00	$5.00	$35.00	$25.00

Nolichucky

(Pictures 102-A thru 102-E)

Nolichucky is a small to medium sized point with expanded hafting area. Average length is between 1-1/4 inches and about 2 inches long and around 5/8 inch in width.

The cross-section is biconvex. The distal end is very acute. Blade edges may either be straight or excurvate. The sides of the hafting area is incurvate, usually expanded at the basal edge forming expanded pointed or rounded auricles. Basal edge is usually thinned and incurvate but may be straight. Sides of hafting area are occasionally ground on the edges.

Faces of the blade and hafting area are shaped by deep to shallow random flaking. All edges show fine retouch flaking.

The type gets its name from the Nolichucky River, where the type was originally defined by Kneberg (and Lewis?) in 1957. There is sparse distribution through north Alabama of Nolichucky points with a suggested occurrence during late Woodland. Examples from the Nolichucky River-Camp Creek area sites are estimated at 2,000 years before present, or early Woodland period. This type is often found on surface sites with the Greenville, and Camp Creek types.

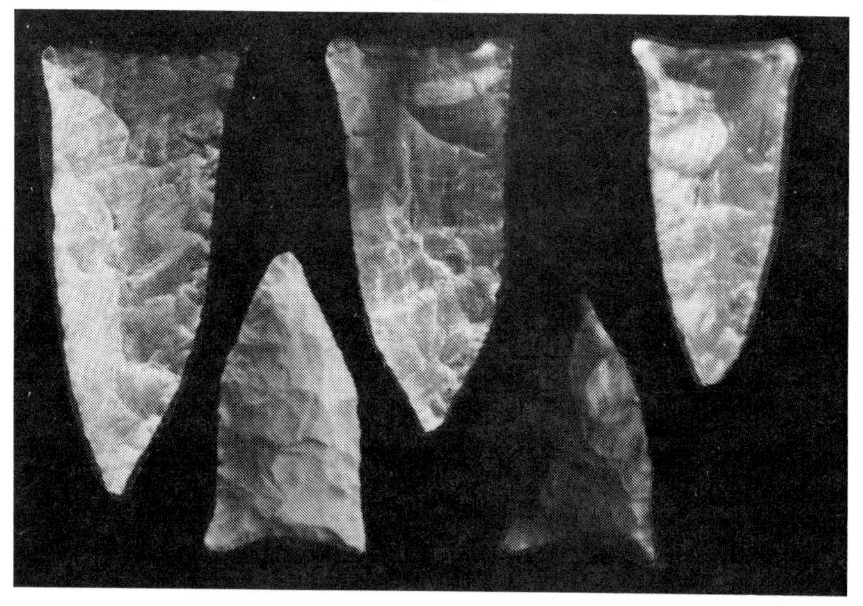

102-A - $75.00 102-B - $45.00 102-C - $40.00
102-D - $30.00 102-E - $20.00

Osceola

(Picture 103-A)

Osceola is a large, deeply inside notched point with parallel blade edges. Average lengths run from 3 inches to 9 inches long and averages 1 inch to 1-1/2 inches in width.

Cross-section is flattened on the faces. Blade edges are usually straight and nearly parallel from above the notches until near the distal end, abruptly going into an acute point. In some areas, blades may show some slight serrations. Notches in the side edges near base form the hafting area, leaving it either rounded or squared on the stem sides with an incurvate base or rarely straight basal edge. Sides of stem are hollow notches, with inside of notches and basal edge usually ground. Some examples have a notched appearance on base.

The blade and hafting area faces were shaped using rather large, but well controlled flaking, followed by long, shallow, random flaking from edge to near center of the blade. This type was typically very well worked.

The type is named after the Osceola site in Wisconsin where the type was noted. The Otter Creek point is somewhat similar to this type. Its estimated age is hedging in the Archaic period. The type is said to be associated with the Old Copper culture which has yielded carbon dates between 5,000 to 7,000 years B.P. Alabama examples are found on predominately early Archaic sites and appears related to the Big Sandy group.

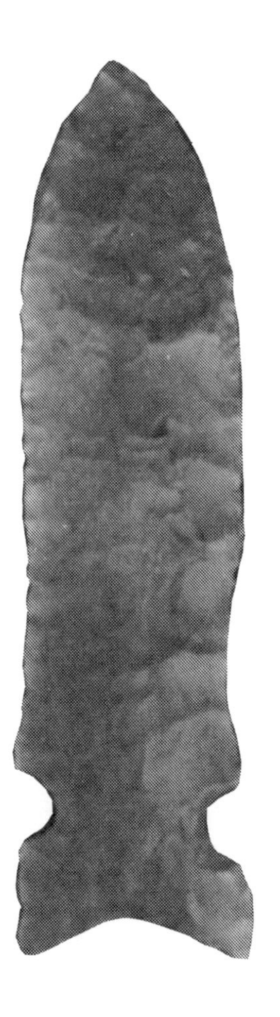

103-A - $800.00

Osceola Greenbrier

(Pictures 104-A thru 104-C)

In form this type resembles a Big Sandy with a fluting nipple left on the base. Originally believed to exist only in north central Florida and southern Georgia, this type has been found as far north as Kentucky.

Notches and all basal edges show grinding. Occasional examples show small flutes or long thinning flakes. Flaking and manufacturing techniques indicate an early Archaic placement.

104-C
$75.00

104-A - $250.00 **104-B - $175.00**

Paint Rock Valley

(Pictures 105-A thru 105-D)

The Paint Rock Valley is a medium sized, triangular point with a broad incurvate base and excurvate blade. Average length for this type is between 1-1/4 inches and 2-1/4 inches long, around 1 inch to 1-1/2 inches wide.

(continued on next page)

The cross-section is biconvex. Distal end is generally acute but may be broad. Blade edges are usually excurvate with no noticeable break between hafting area and blade, in some examples blade may be straight. The sides of the hafting area may contract slightly at the base. Basal edge is most often excurvate but rarely has a nearly straight base. Base is usually thinned or beveled.

Broad random flaking of variable depth was used to shape faces of the blade and hafting area. Fairly long secondary flakes from the blade edges toward center of blade were removed to complete the shape of the blade. Fine retouch is only occasional and slight. Basal edge usually exhibits thinning using short flakes that sometimes leave the edge beveled.

This type was named for Paint Rock River Valley in Jackson County, Alabama, where points of this type were first noted. There is a noted similarity of this type to several western point types which are believed to be 4,000 to 6,000 years old. In north Alabama this type is found in levels and surface collections with points such as Cumberland Quad, Dalton, Wheeler, Big Sandy, Morrow Mountain, White Springs, Crawford Creek and related tools and bone point. These associations suggest an age from transitional Paleo to early Archaic times.

105-A - $75.00

105-B - $60.00 **105-C - $45.00** **105-D - $35.00**

Palmer

(Pictures 106-A thru 106-D)

Palmer is a small corner notched point with straight base and strong serrations. Average length is from 1-1/8 inches to about 2-1/4 inches with 1-3/4 inches being medium and from 1/2 inch to 1 inch in width at shoulders.

Cross-section is biconvex. The blade edges are most often straight but may be slightly incurvate or excurvate. Examples are usually serrated, in some cases quite deeply, along the blade edges. Shoulders are barbed. The hafting area is corner notched forming an expanded stem with usually incurvate sides, with a straight or occasionally incurvate or excurvate basal edge. Basal edge is thinned and ground.

Blade and hafting area faces were pressure flaked from proper sized prismatic flakes. After blade was shaped a very regular deep secondary flaking on the blade left serrations on the edges with flake scars often meeting or overlapping in the middle of the blade face. Notches were placed by removal of deep broad flakes into the corners of the blade.

(continued on next page)

The type was named for points first noted at Hardaway site in Piedmont, North Carolina. The point has limited distribution, spread over most of the southeast and up the Atlantic seaboard as far as New England. Despite this wide area it is spread over, Palmer is a scarce type. An estimate of about 8,000 years old is suggested.

106-A-$125.00 106-B-$150.00 106-C-$45.00 106-D-$45.00

Pedernalis

(Pictures 107-A thru 107-C)

Pedernalis point is medium to large with a bifurcated stem. Length varies widely from 1-1/8 inches to over 5 inches long with a width of around 1-1/4 inches.

Cross-section is either flattened or biconvex. Blade is usually straight or excurvate, but in some cases will be incurvate or recurvate. Distal end may be acute or occasionally comes to needle-like narrow end. Shoulders may be either horizontal or barbed. Base side edges are straight or at times slightly excurvate. Stem may be contracted in some examples. Basal edge is usually deeply incurvate and may be thinned or slightly beveled. Basal grinding is not commonly present.

Blade and hafting area is shaped by removal of broad, shallow flakes, with deeper, shorter flakes along all edges. Large broad flakes were usually removed to form the shoulders. Basal edge may be thinned by removal of one large flake or several thin ones from the basal edge.

(continued on next page)

This type is named for the Pedernalis River in Texas where the point was first recognized. The type is considered fairly common throughout central Texas were it is estimated to be from 6,000 to possibly as recent as 2,000 years before present. An early Woodland to late Archaic span is suggested.

107-A - $450.00 **107-B - $500.00** **107-C - $75.00**

Pickwick

(Pictures 108-A thru 108-F)

Pickwick is a medium to large point with expanded shoulders and a recurvate blade. Length varies from 2-1/2 inches to over 5 inches but will average around 3-1/4 inches long and from 1-1/2 inches to nearly 2 inches in width.

(continued on next page)

Cross-section is biconvex. Blade edges are recurvate often with fine serrations. Distal end is acute. Shoulders are always expanded and usually tapered but may be horizontal. A large portion have asymetrical shoulders. The stem is fairly thick and is usually tapered but rarely may be straight. Side edges tend to be incurvate and are sometimes ground. Basal edge is either excurvate or straight. Basal edge is only rarely ground.

Shaping of the blade and stem was carried out be removal of broad, shallow flakes with a short, deep, regular retouch. Blade retouch sometimes resulted in fine serrations. Only light retouch is used on the stem edges. Material used is from local sources.

The type is named for the area in which they were first noted in the Pickwick basin of the Tennessee River Valley. Excavated examples suggest an emerging during middle Archaic and continuing and increasing into the late Archaic period.

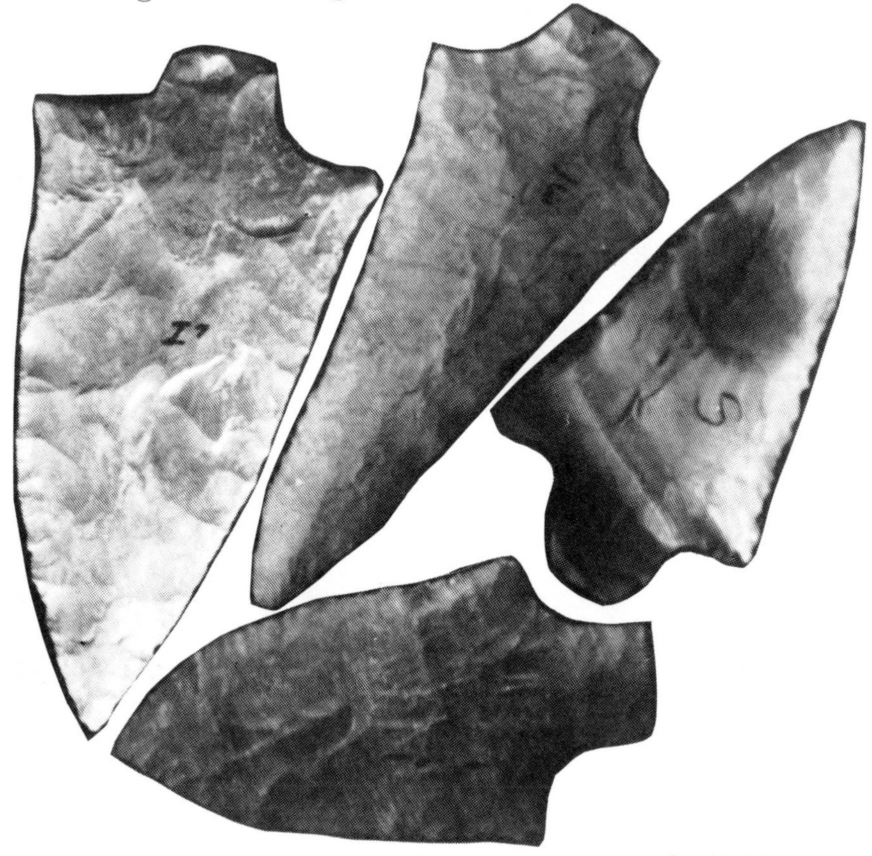

108-A - $400.00 108-B - $100.00 108-C - $250.00
108-D - $75.00 Pickwick Tri-Color

108-E - $1,000.00

108-F - $600.00

Pine Tree

(Pictures 109-A thru 109-E)

Pine Tree is a medium-sized, side notched point, usually serrated with expanded shoulders. Average length is between 2 inches and 2-1/2 inches long and from 1 inch to 1-1/4 inches in width.

Cross-section may be biconvex or rarely flattened. Blade edges are recurvate and serrated. Shoulders are tapered, narrow, and expanded. Distal end is acute. The side notches in the hafting area forms an expanded stem, with incurvate side edges and an incurvate basal edge. Basal edge is thinned.

Blade and stem were shaped by broad, shallow flaking. Retouching with collateral or well spaced random flakes from the blade edges, with flakes often meeting in middle of face, left a slight median ridge. This retouch usually left well spaced serration along blade edges. Side notches are formed when one large flake or several smaller flakes was struck from sides of blade near base. Edges of base are finely retouched.

The type is named for points found around the Cambron site known as the Pine Tree site in Limestone County, Alabama. The type appears on pre-shell-mound Archaic site in relation with early Archaic tools and point types. This indicates an early Archaic cultural association.

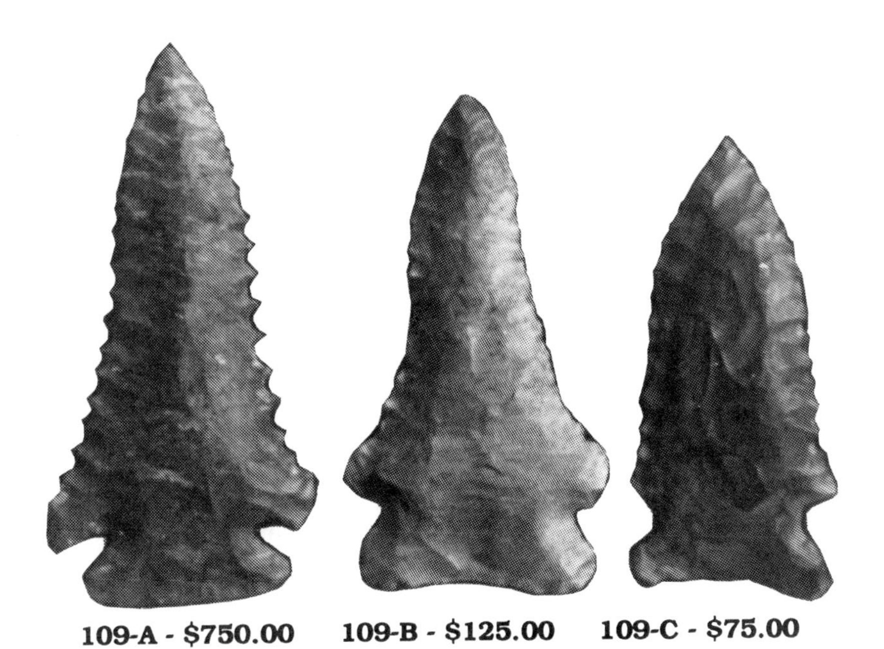

109-A - $750.00 **109-B - $125.00** **109-C - $75.00**

109-D - $2,000.00
109-E - $750.00

Pine Tree Corner Notched

(Pictures 110-A thru 110-I)

This is a medium sized point with expanded shoulders, side notches and serrated blade edge. Average length is from 1-1/2 inches to 2-3/4 inches long and from about 7/8 inch to 1-1/4 inches in width.

Cross-section is biconvex or rarely may be flattened. Blade edges are generally incurvate but in rare cases may be straight and are always serrated. The distal end is acute. Shoulders are tapered inversely or occasionally horizontal. The barbs are usually expanded. Hafting area is corner notched with an expanded stem with straight sides. The basal edge is usually straight, but may also be excurvate, or rarely, incurvate. Basal edge is thinned with light basal grinding evident on most points.

Shaping of the blade and hafting area utilized random, fairly shallow flaking. The point was then retouched along the blade edge with collateral flaking that resulted in serrations on the blade edge.

(continued on next page)

Retouch was worked into the blade to form the expanded barbs. On some, more large flakes were removed from corners to form the notches, then all edges of the base were retouched.

The type name is derived from Cambron's Pine Tree site in Limestone County, where early examples of the type were recognized. The type is a variant of the Pine Tree type. Workmanship seems to indicate a slightly older occurrence than the Pine Tree. This type is typically found on pre-shell-mound early Archaic sites in Alabama and southern Tennessee.

110-A - $1,200.00 110-B - $750.00 110-C - $500.00

110-D - $700.00   110-E - $900.00   110-F - $75.00

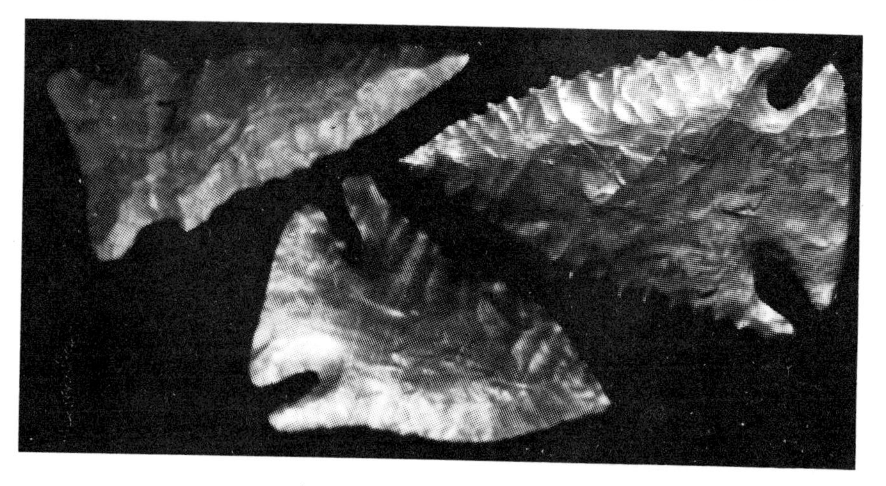

110-G - $50.00   110-I - $125.00
110-H - 250.00
(This point (H) stolen, if seen, contact author!)

Plevna

(Pictures 111-A thru 111-H)

Plevna is a medium to large, corner notched point with beveled blade edge on one side of each face. Average length ranges between 1-3/4 inches and 3-3/4 inches long and between 1 inch and 1-3/8 inches in width. Points of this type exist over 4 inches long, but are very rare.

The Plevna has a rhomboid cross-section. Blade is generally straight but on some examples may be incurvate or excurvate. Blade edges are beveled sharply on one side of each face and may in some cases be serrated. Shoulders may be inversely tapered or horizontal. Shoulder barbs, if present will be expanded. Distal end is acute. Hafting area has deep, narrow, corner notches leaving an expanded stem with an excurvate basal edge. Stem is usually well thinned and ground on basal edge.

Broad, shallow, random flaking was used to shape faces of the blade and hafting area. Blade edges on one side of each face are steeply flaked along edge resulting in the beveled edges. This retouch also left serrations on some points. Plevna was notched at widest point in the blade near the base by removing a large deep flake on each side of each corner. This produced a thinned area which was then shaped by taking short fine flakes, leaving a long, narrow, corner notch. The hafting area was thinned first with fairly large, shallow flakes and then finely rechipped on basal edge.

The type is named from points on the Plevna site, in Madison County in north Alabama. This site included a number of other early Archaic types and tools along with numerous representatives of Plevna types. Antiquity in north Alabama is some time prior to 5,000 years old, during the early Archaic period.

(points on following pages)

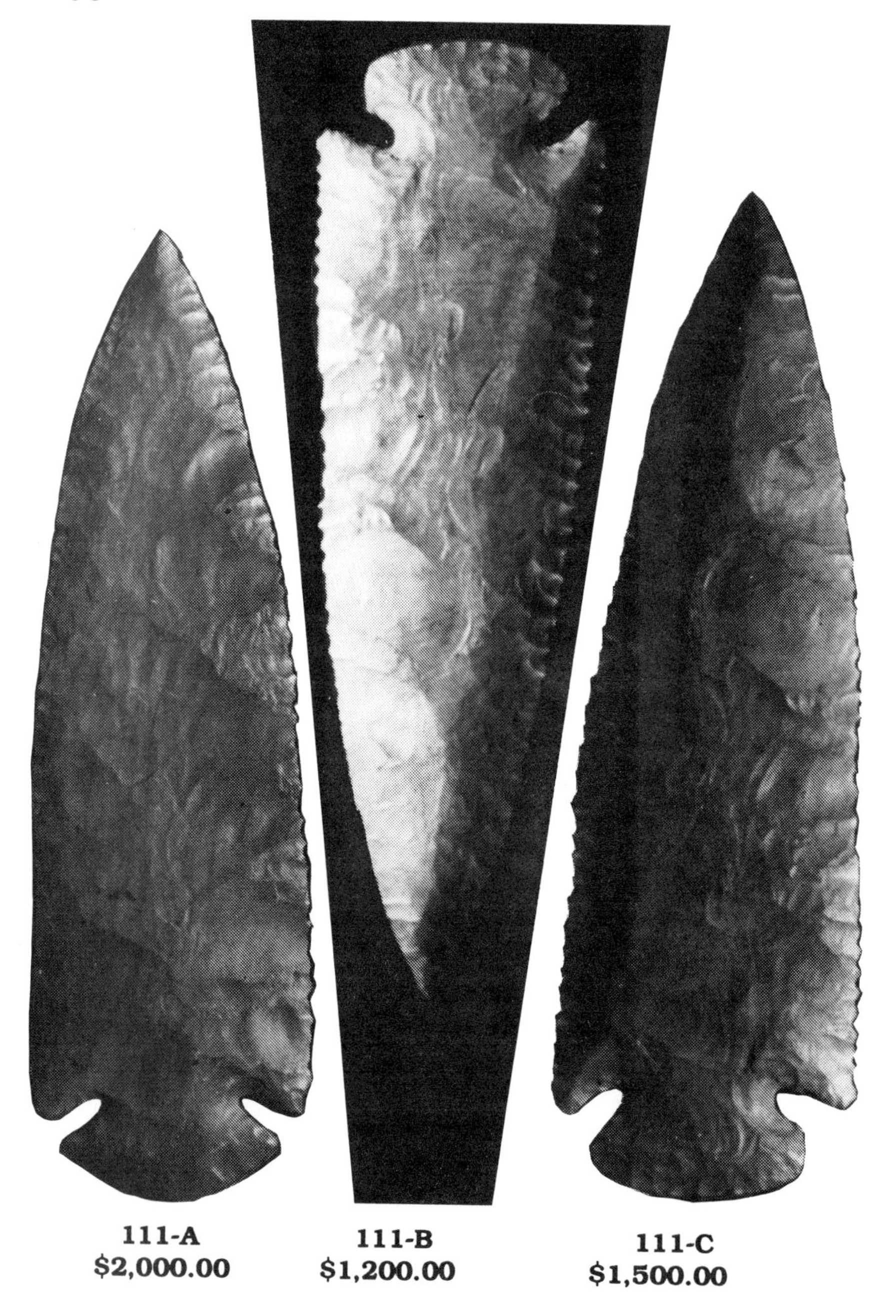

111-A
$2,000.00

111-B
$1,200.00

111-C
$1,500.00

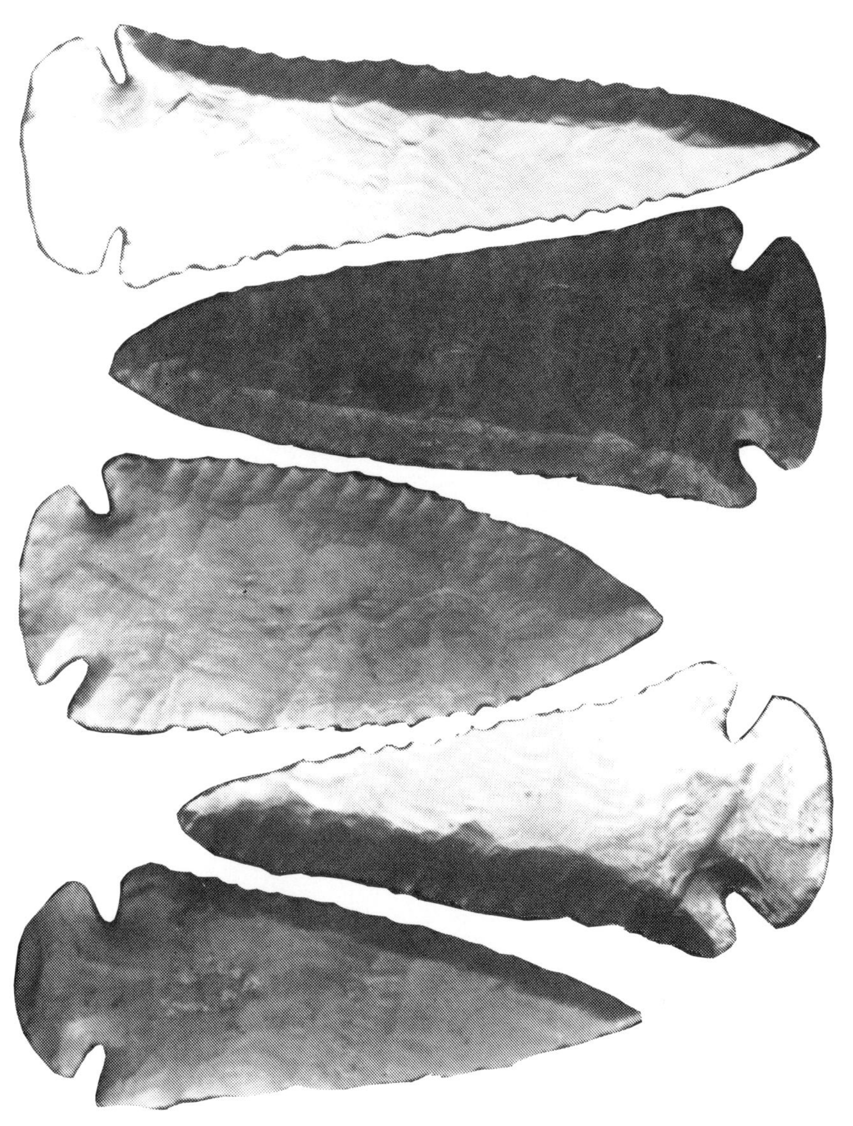

111-D - $1,500.00
111-E - $400.00
111-F - $500.00
111-G - $750.00
111-H - $400.00

Putnam

(Picture 112-A)

A medium size point type found in the coastal plains and coastal region of the southeastern United States. The blade edges are recurvate sometimes straight. Shoulders are weak with no definite junction between stem on shoulders. Basal edges are rounded. Age: late Archaic to early Gulf formational.

112-A

$75.00

Quad

(Pictures 113-A thru 113-H)

Quad is a medium sized, wide point with deeply incurvate base with an auriculate expanded-rounded hafting area. Average length is from a little over 2 inches to around 3-1/2 inches, though longer points are known, they are rare. Width averages between 3/4 inch and 1 inch.

Cross-section is either biconvex or flattened. The distal end is acute. Blade edges are excurvate. Hafting area has incurvate sides with incurvate base. Base is auriculate with auricles being expanded-rounded. All edges of the haft are usually well ground. Basal edge may be thinned or fluted.

Blade and hafting area is usually shaped by shallow random flaking but may be collateral. Secondary flaking is present on all edges, using short fairly deep flaking to finish the point. On fluted examples flute may be singular or multiple, usually short and fairly broad as on many Clovis points.

(continued on next page)

The type, was named after points found in or around the Quad site in Limestone County, Alabama. Distribution is scattered from Ohio to Alabama usually appearing on transitional Paleo sites in conjunction with such types as Big Sandy I, Wheeler, Cumberland, Dalton, Beaver Lake and various uniface scrapers and tools. Bell, in describing the type, indicated a supposed antiquity of between 1,000 and 6,000 years ago. From information gained from excavated and surface finds, a transitional Paleo association appears certain.

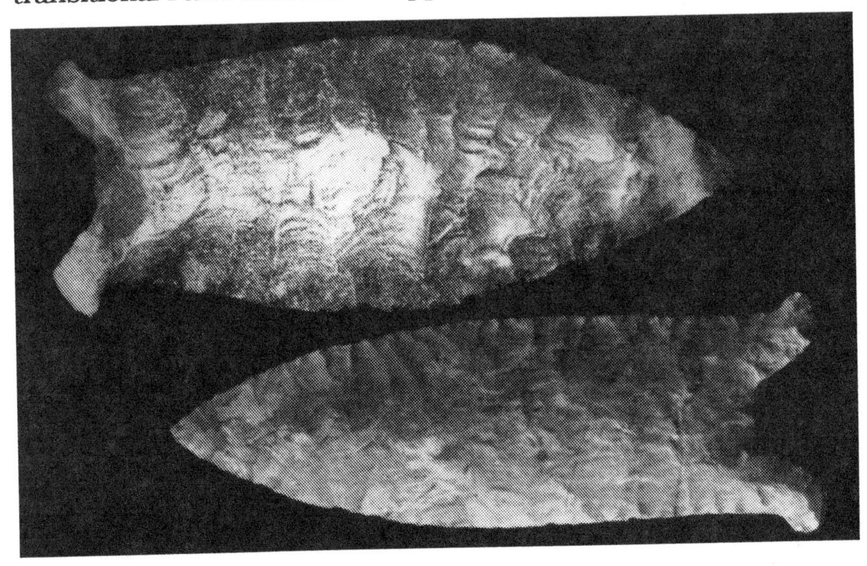

113-A - $2,500.00
113-B - $2,000.00

113-C - $300.00
113-D - $250.00

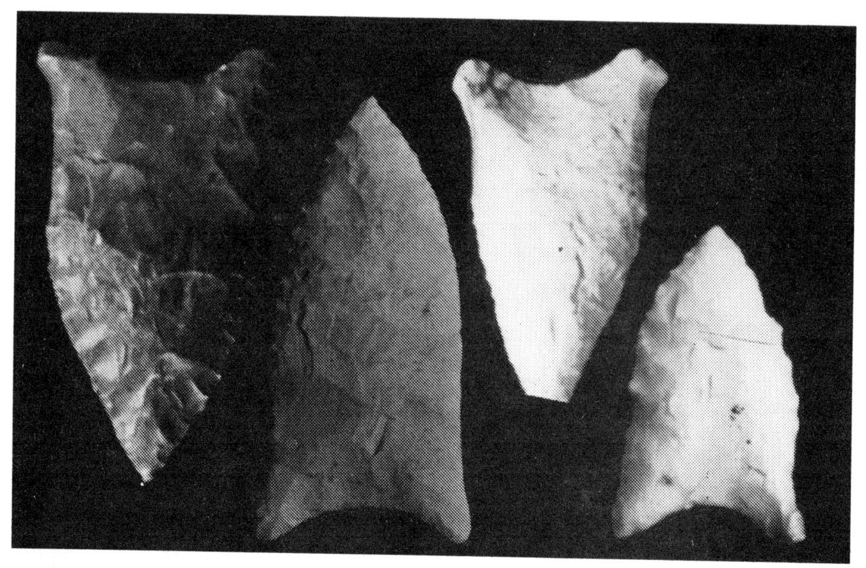

113-E-$175.00 113-F-$175.00 113-G-$35.00 113-H-$75.00

Redstone

(Pictures 114-A thru 114-F)

A medium to large, triangular, fluted point. Usually exhibits multiple fluting. The auricles are always the widest part of the point. Average lengths are 2 inches to 7 inches and 1-1/2 inches to 2 inches in width.

Examples larger than 3 inches are almost non-existent. As usual flea market examples are almost always larger than three inches and plentiful.

This is the rarest of the fluted points. Fluting is multiple; two short flutes were stuck off each side of the platform then a large center flute was stuck by direct percussion down the blade between the two side flutes.

Where Clovis and Cumberland points were biconvex to medium ridged on each side, Redstone had one flattened face and one medium ridged face. The flattened face was stuck first then the medium ridged face last. The basal edge was thinned and ground. Redstone points usually have the finest retouching of the three Paleo types.

The Redstone was named after the Redstone Arsenal in Madison County, Alabama by Mahan in 1954.

(points on next page)

114-A - $$5,000.00
114-B - $3,000.00
114-C-$2,000 114-D-$1,000 114-E-$500 114-F-$500

Russell Cave

(Pictures 115-A thru 115-E)

Russell Cave is a medium sized point with expanded stem and serrated, straight, blade edges. Length averages from 1-3/4 inches to 2-1/2 inches long and from 1 inch to 3/4 inch in width.

Cross-section is biconvex. The blade is nearly always straight but may occasionally be excurvate. Blade edges exhibit shallow serrations. Shoulders are tapered, the point has an acute distal end. The stem has incurvate sides with a straight basal edge. All sides of base edges are usually ground. Basal edge may be beveled narrowly.

Shaping the blade used shallow, broad, random flaking, with short, deep flakes finishing the blade with regular shallow serrations on edges. Serrations are raised by removing the deep flakes from opposite sides of edge, leaving very sharp points. Steep flaking on sides of stem left a somewhat expanded base. Thinning flakes were usually removed from basal edge leaving fairly broad flake scars. In some examples of this type the basal edge may be slightly beveled by the removal of short steep flakes.

The type was named for Russell Cave in Jackson County, Alabama by Cambron. The site, where the seven points used to originally describe the type, contained charcoal in context. The result was a carbon date of between about 7,000 to 9,000 years old. For this reason the point appears to be of early Archaic origin. Collectors in Kentucky and Tennessee consider the Russell Cave to be a type of Harpeth River and call it Harpeth River #2. In South Alabama and Georgia, local collectors call the Russell Cave point Boggy Branch #2.

115-A - $375.00

115-B-$100.00 **115-C-$75.00** **115-D-$100.00**

San Patrice

(Pictures 116-A thru 116-E)

A Dalton shaped point found in the lower Mississippi River Valley and into southern Alabama. It occurs as several variants. Base exhibits basal thinning or fluting. It is often serrated. Age is from Transitional Paleo to early Archaic.

116-A - $450.00 **116-B - $500.00**

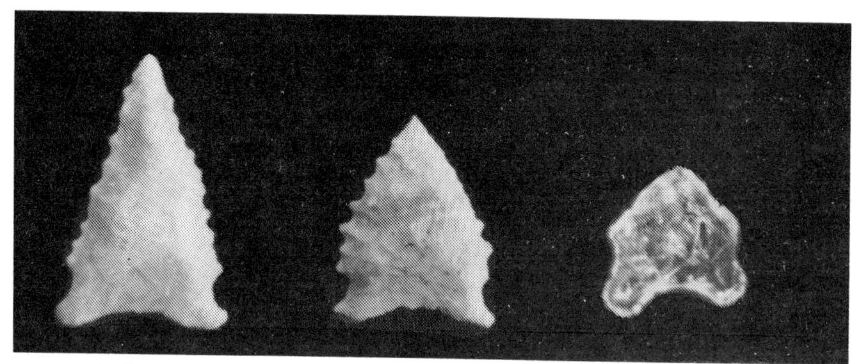

116-C - $100.00 **116-D - $75.00** **116-E - $25.00**

Sand Mountain

(Pictures A thru G on following page)

Sand Mountain is a small triangular point with serrated blade and incurvate base. Average length is from 1 inch to 1-1/4 inches long and around 1/2 inch in width at base.

Cross-section is generally biconvex, but is flattened in a few examples. The blade tends to be straight but may be incurvate or excurvate. Blade edges exhibit distinct serrations. The distal end may either be acute or occasionally acuminate.

Basal edge is incurvate and pointed at junction of blade and basal edge. Broad, fairly deep, pressure flaking removed from alternated faces of the blade shaped and at the same time left the prominent serrations on blade. Basal edge has fairly broad flakes removed to thin base. A large number of this type were made from quartzite but other materials found locally may be used.

The point is named for sites on Sand Mountain in northern Alabama. This type has been associated with several varieties of pottery such as Weeden Island and Autauga Check Stamped. This, along with other evidence from sites in the Tennessee River Valley, seems to indicate an emergence in the late Woodland period existing into the Mississippian period.

(points on next page)

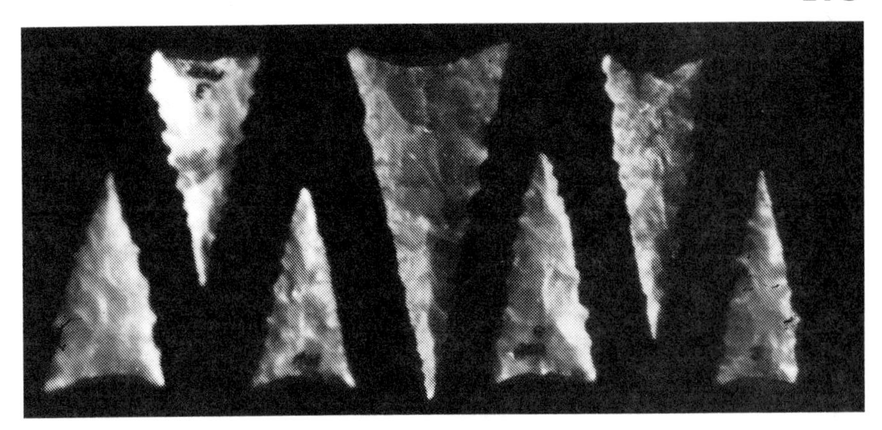

LEFT TO RIGHT:

A	B	C	D	E	F	G
$25.00	$15.00	$5.00	$50.00	$15.00	$30.00	$20.00

Santa Fe

(Pictures 118-A thru 118-B)

Triangular shaped point, often shows basal grinding related to the Tallahassee points and is sometimes found associated with Fiber Tempered Pottery. Similar examples are found in north Alabama and are thought to be associated with the Wheeler Complex.

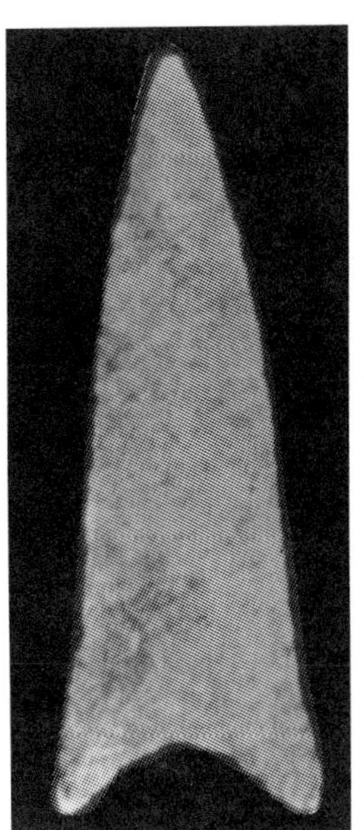

118-A
$350.00

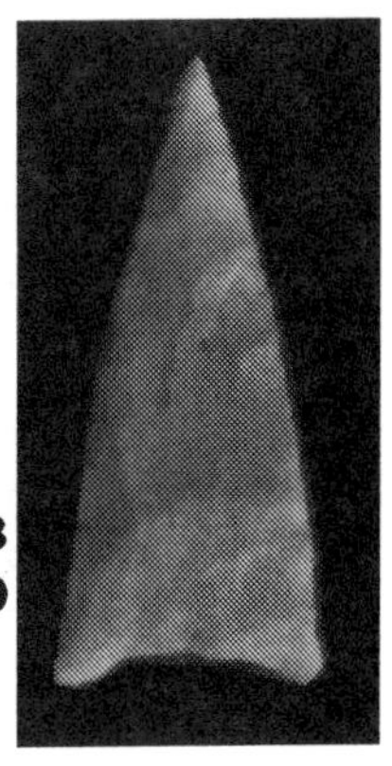

118-B
$75.00

Savannah River

(Pictures 119-A thru 119-B)

Savannah River is a medium to large point with fairly broad stem, usually made from quartzite. Average length is from 2-3/4 inches to 6-3/4 inches long and from 1-1/2 inches to 2-3/4 inches in width at shoulders.

Cross-section tends to be biconvex but may be flattened or on rare occasions, may be plano-convex. The blade edges are excurvate. In some examples blade edges may approach parallel from shoulders to about 1/3 the length of the blade. The shoulders are most often tapered but may be straight. The distal end is acute. The stem sides are straight with basal edge being either incurvate or straight. Stem may be straight or tapered. The basal edge is usually thinned.

Shaping of the blade and hafting area utilizes broad shallow flaking followed by fine retouch on all sides.

The type was named for the Savannah River Focus of the Archaic period around the Piedmont of North Carolina. Distribution is known to occur in North Carolina, Georgia, and Alabama almost exclusively. Some examples have in the past been referred to as Appalachian-Stemmed, that better conform to this type. Evidence points to a late Archaic to early Woodland association.

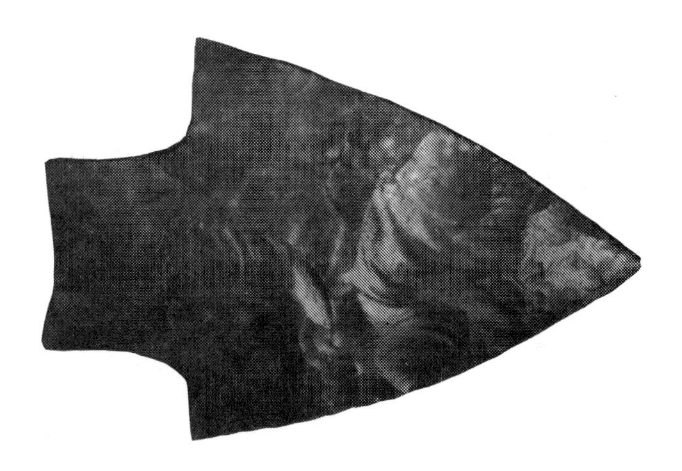

119-A - $500.00

119-B - $175.00

Sedalia

(Pictures 120-A thru 120-C)

A large size lanceolate point. Cross-section tends to be thick. Widest part is usually mid-distance from tip to base. Sides are excurvate and slightly recurvate near the base. This probably denotes the hafting area. Flaking is random with good retouching. Age: early Archaic.

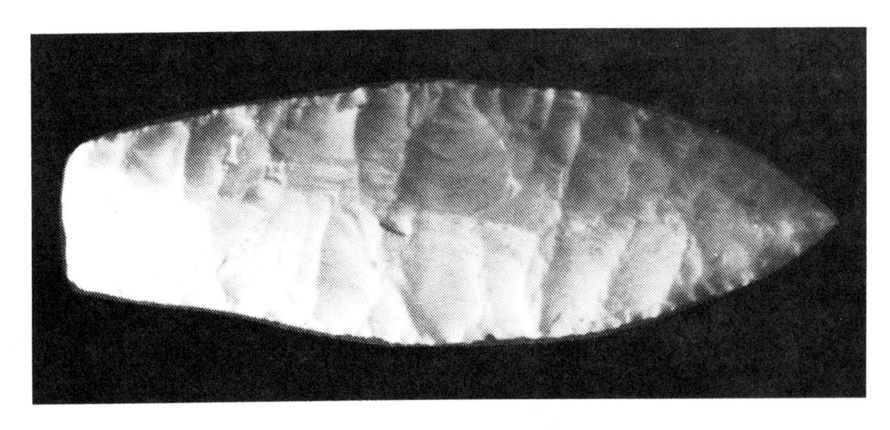

120-A - $150.00

120-B - $75.00
120-C - $200.00

Shoals Creek

(Pictures 121-A thru 121-F)

A medium to large stemmed point. Blade edges are straight to slightly incurvate. Shoulders are straight. Stem is usually straight but may contract slightly and is sometimes unfinished with a broken or fractured appearance. Flaking is well controlled and random which approaches collateral and often results in a median ridge. Blade edges are finely retouched which results in good serration. It is found on sites with Elora, Little Bear Creek, and other late Archaic and early Woodland point.

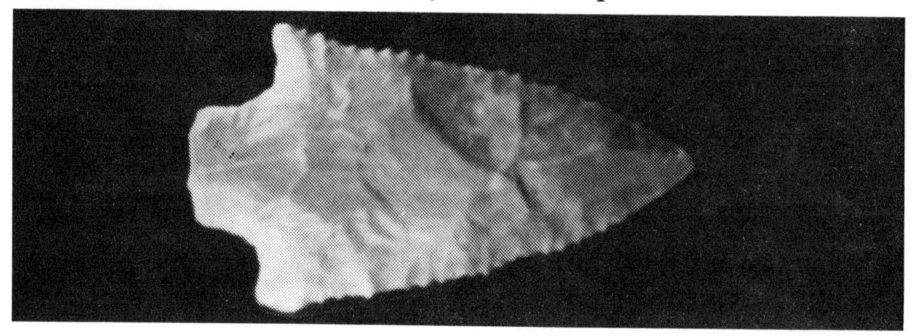

121-A - $50.00

121-B - $250.00 121-C - $300.00 121-D- $100.00

121-E - $75.00 121-F- $75.00

Simpson

(Pictures 122-A thru 122-I)

A medium to large auriculate point or blade. The hafting area is constricted. The blade edges are recurvate and the base is expanding auriculate. Widest point of blade is one third to one half of the way from the tip to the base. All examples not fluted are thinned. The hafting area is ground. Has both Clovis and Cumberland characteristics. Collectors in the Tennessee River area refer to it as a Cloverland Point. Age is late Paleo.

122-A - $2,500.00
Rare Mustache Simpson. Only Intact Example Known.

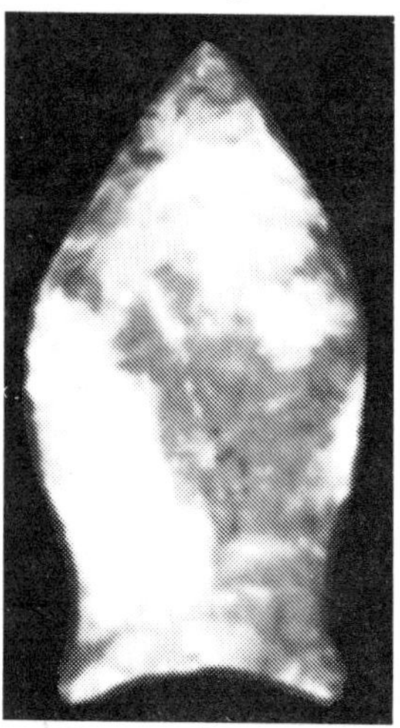

122-B - $200.00

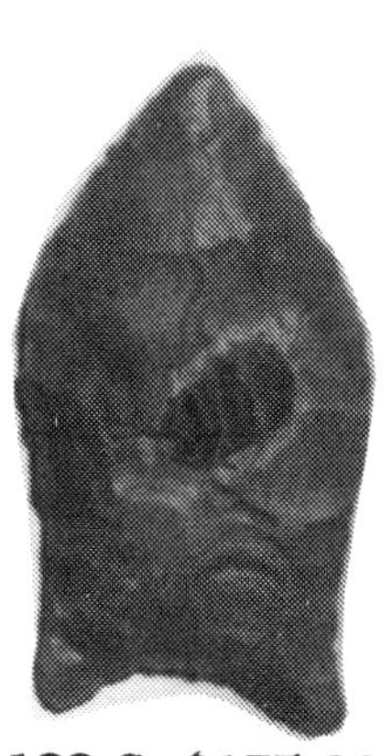

122-C -$175.00

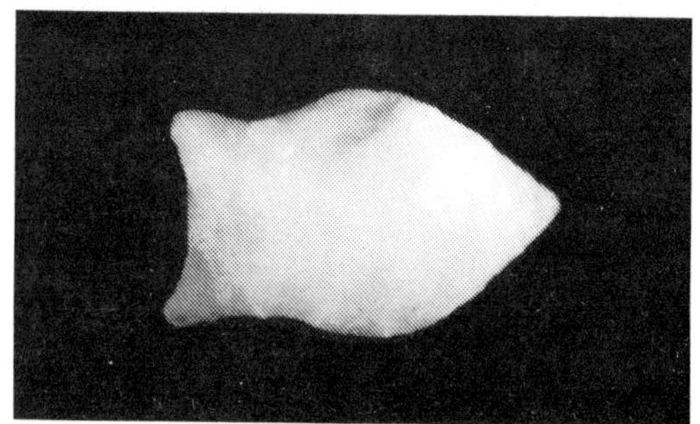

122-D- $100.00

122-E - $10,000.00
Best Example Known

122-F - $3,500.00 122-G- $750.00 122-H - $500.00
122-I - $350.00

Smithsonian

(Pictures 123-A thru 123-E)

Smithsonian is a medium to large point with straight blade and stemmed base. Average length is from 2-1/4 inches to 3-1/4 inches long and averages 1-1/4 inches in width at shoulders.

Cross-section is generally biconvex but is occasionally flattened. The point is finely worked on blade edges leaving fine even serration. Blade edges are usually straight but may be excurvate or on some examples, one blade edge is excurvate with the other blade straight. The distal end is acute. Shoulders are usually barbed but may be tapered. Stem and base edges are always straight with thinned and lightly ground basal edge. The stem sides are often thinned, sometimes being lightly ground.

Broad, shallow, random flaking shaped the blade and hafting area faces. Blade edges were then finely worked by removal of very well controlled, somewhat shallow, flakes, leaving a characteristic fine shallow serration. Thinning on basal edge is fairly broad shallow flakes. Thinning flakes on stem edge are usually thinner, shorter, more steeply oriented.

The type was named by Ralph Allen for site 41 at Smithsonian on the Tennessee River in Alabama. The area of distribution at this time appears to be restricted to sites near the Tennessee River where they are found in association with points and tools that indicate a late Archaic age.

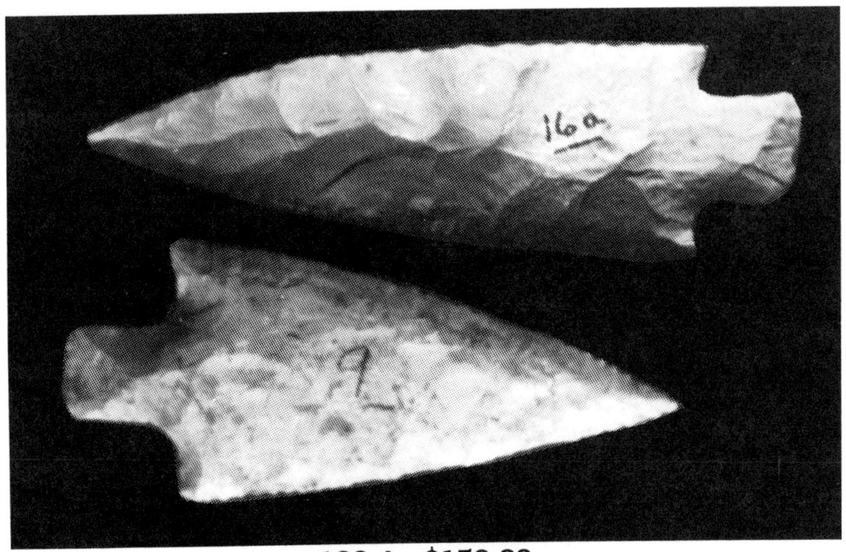

123-A - $150.00
123-B - $125.00

123-C - $65.00 **123-D - $100.00** **123-E - $75.00**

South Prong Creek

This type is a large, broad shouldered, stemmed point with part of the blade serrated but not to the point. Average length is from about 1-3/4 inches to 4 inches long and 1-3/4 inches to 2-1/4 inches in width.

Cross-section is usually straight, but some examples may be slightly excurvate or recurvate. Serrations are an important characteristic in identifying this type. Serrations are located above the shoulders and stops at a point about 3/4 of the distance from base to distal leaving one quarter at the distal end without serrations. There is usually a short gap between shoulder and beginning of serrations. The distal end is acute. Stem is short with straight sides and either straight or excurvate on the basal edge. Stem sides and basal edges are usually thinned and lightly ground.

(continued on next page)

Blade faces and hafting area were shaped by well executed, shallow, random flaking. The blade was serrated by removal of small, deep, fairly short flakes. This created rather deep serrations and sometimes resulted in a bevel in that area. Fairly large flakes were removed from corners of blade to form stem. The shoulder as well as the stem sides was retouched.

The type was named for South Prong Creek, location of the site where type is found in Richmond County, Georgia. The type does not appear to be widely distributed and little hard evidence exists as to origins. On Lewis site 606, the site where type is named for, it was found in association with numerous transitional Paleo and Archaic type with only a few Woodland to Mississippian artifacts. From structure, workmanship, and proximity to other types considered, indication of an early to middle Archaic occurrence possibly beginning around 5,000 years old or before.

124-A - $50.00 **124-B - $25.00**

Southern Hardin

(Pictures 125-A thru 125-J)

Southern Hardin is a medium size, corner notched point. Similar in flaking characteristics to other Hardins. However, this type has a rounder or oval basal edge and is much smaller on average than the northern varieties. Cross section is thin. Notches have the appearance of being the points of a crescent moon. One of the best made of all point types.

(points on following page)

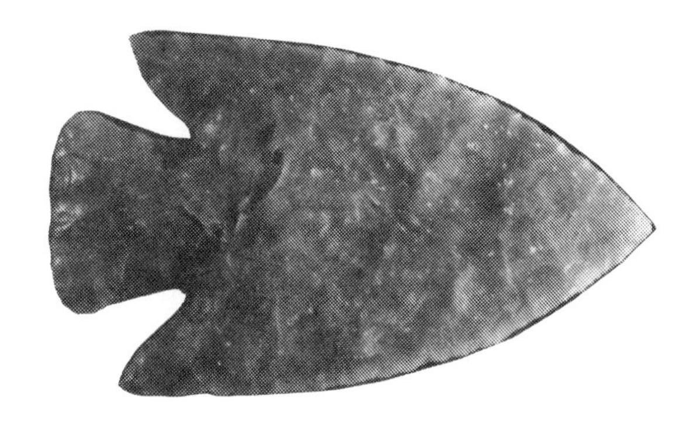

125-A - $500.00

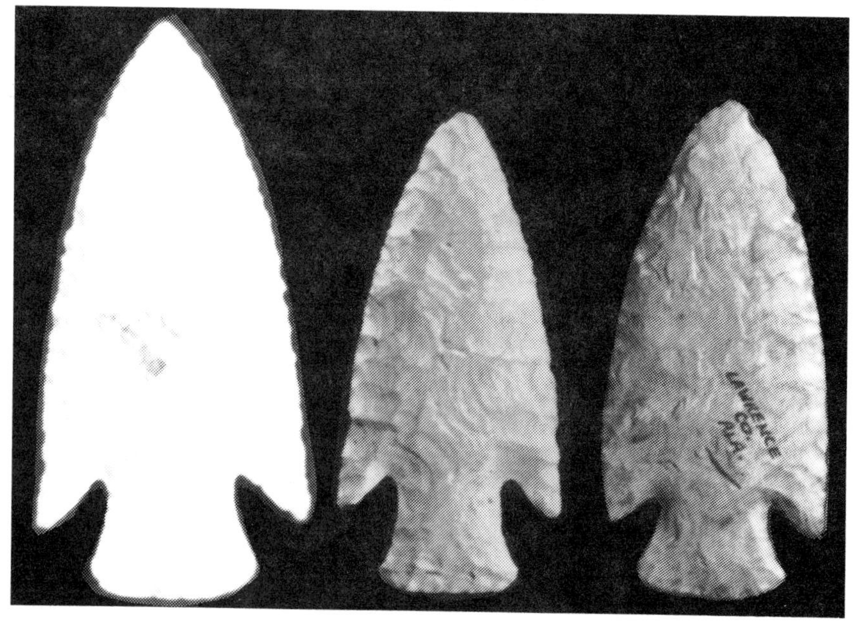

125-B - $750.00 125-C - $300.00 125-D - $150.00

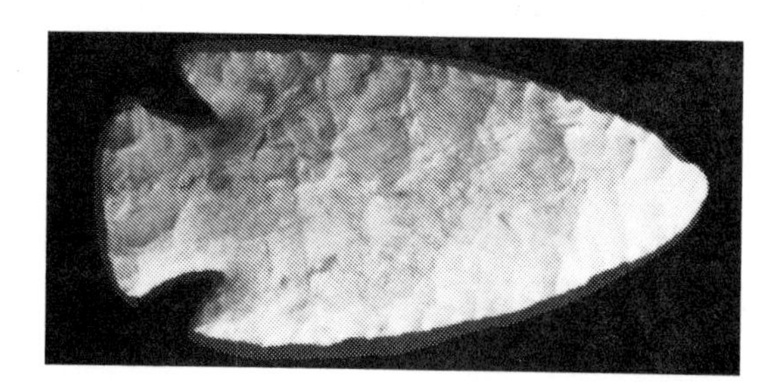

125-E
$150.00

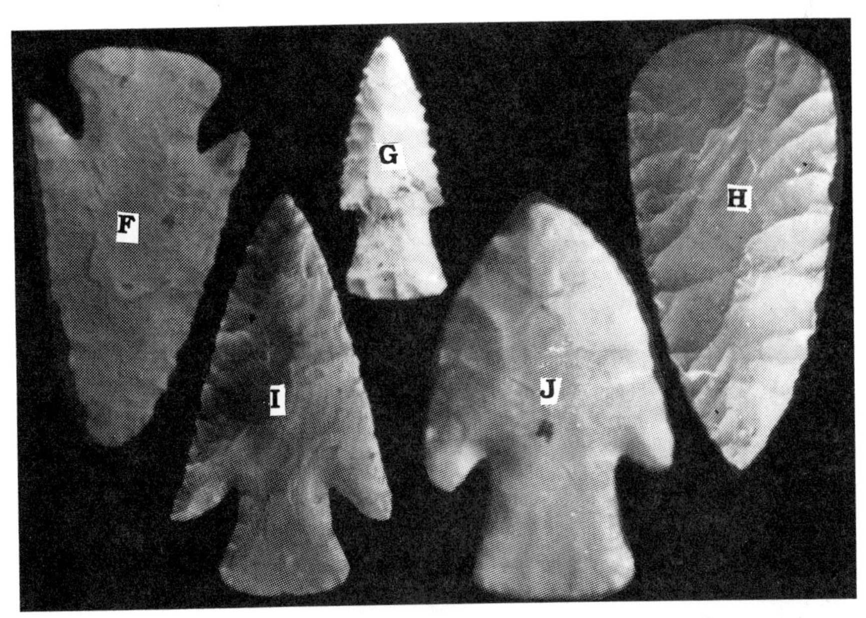

125-F - $45.00 **125-G - $50.00** **125-H - $100.00** **(Preform.)**
125-I - $300.00 **125-J - $45.00**

Spring Creek *(Pictures 126-A thru 126-C)*

Spring Creek is a large, barbed, stemmed point. Blade edges are recurvate, shoulders are strongly barbed. Stem sides expand and basal edge is straight to slightly incurvate. On nearly all examples, one face of the blade is almost flat while the other side is lenticular in cross-section. It is found on late Archaic and early Woodland sites in association with Motley, Little Bear Creek and Elora points. At one time this point was thought to be associated with the Pickwick culture.

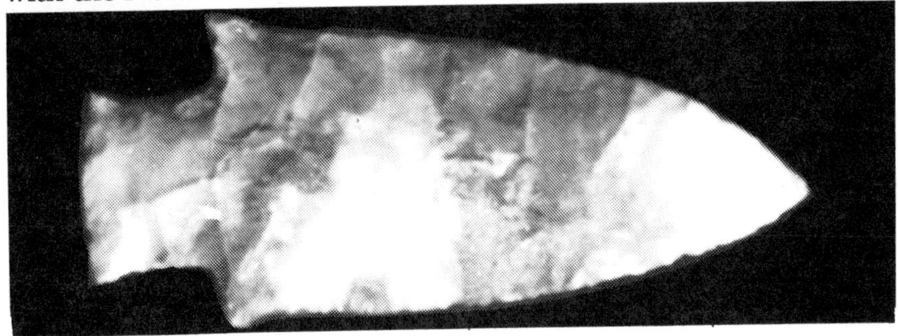

126-A - $800.00 *(Multi-color, orange and gold)*

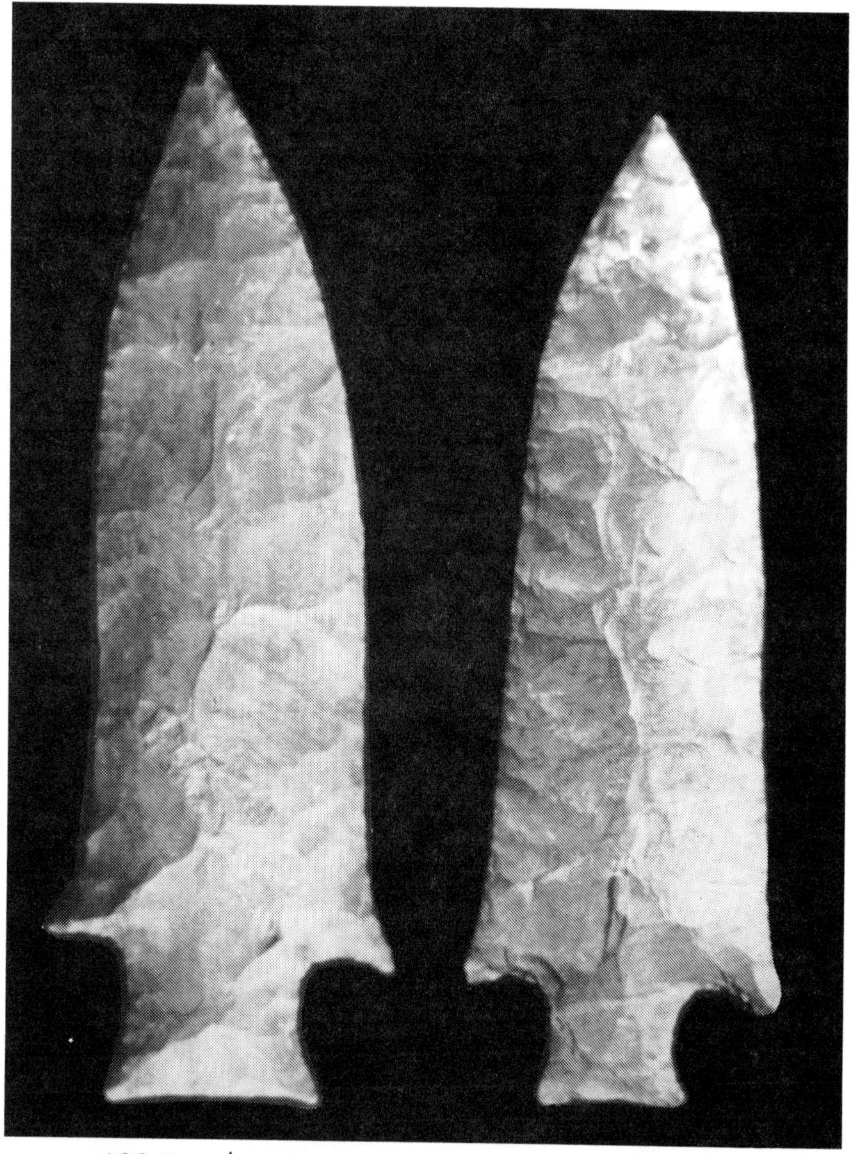

126-B - $1,000.00

126-C - $800.00

Stanfield

(Pictures 127-A thru 127-B)

Stanfield is a medium to large point with somewhat triangular shape and flat base. Average length is from about 2-1/4 inches to 4-3/4 inches long and 3/4 to 1-1/4 inches in width.

Cross-section from distal end to beginning of hafting area is biconvex, but in most cases, the hafting area is flattened. Blade will usually have parallel edges, except in rare instances when it may be excurvate. Distal end is usually acute. Hafting area is not readily noticeable where it joins the blade, the sides are usually parallel with a straight basal edge. Basal edge is usually thinned with some grinding on most examples.

Faces of blade and haft were shaped by removal of deep to shallow, random flaking. From the point at distal end down about 1/3 to 2/3 of the blade length, will be flaked from edge to form a median ridge. Small somewhat broader flakes were removed to shape the hafting area. Short, broad and steeper, deeper flakes were used as a retouch on edges of blade and haft to finish the shape and edge.

The type is named from examples recovered in the excavation of Stanfield-Worley Bluff Shelter in Colbert County, Alabama. Examples of this type have been found in conjunction with Dalton, Big Sandy I, Lerma and uniface tools. Examples from type site was carbon dated 8920+400 and 9640+450 years before present. From the information provided by excavation and related artifacts from surface sites, it can be assumed to be transitional Paleo Indian culture artifact.

127-A - $100.00 **127-B - $50.00**

Stanley

(Pictures 128-A thru 128-B)

Stanley is a medium size triangular blade point with stemmed base. Average length - about 1-3/4 inches long and average width of 1-1/4 inches.

Cross-section may be biconvex, plano-convex, or rarely is flattened. The blade edges are either excurvate, straight, or recurvate. Distal end usually has an angular break from the blade edges to the point. Shoulders are generally horizontal or tapered, on some examples, may be expanded. Distal end is either acute or somewhat apiculate. Hafting area is a straight stem with incurvate or notched basal edge.

Blade and haft faces were shaped by broad random flaking followed by some retouch leaving longer thinner flakes scar on blade. Some examples show straight serrations getting deeper as they get closer to the shoulders. A fine retouch was used to finish the blade and distal area. Corners were removed with fairly deep, wide flaking to form stem. Long thin flakes were removed from basal edge to thin the stem.

The type was named for points off sites around Piedmont, North Carolina, where the point was first noted. Types in excavated sites seem to point to an Archaic origin about 5,000 years ago.

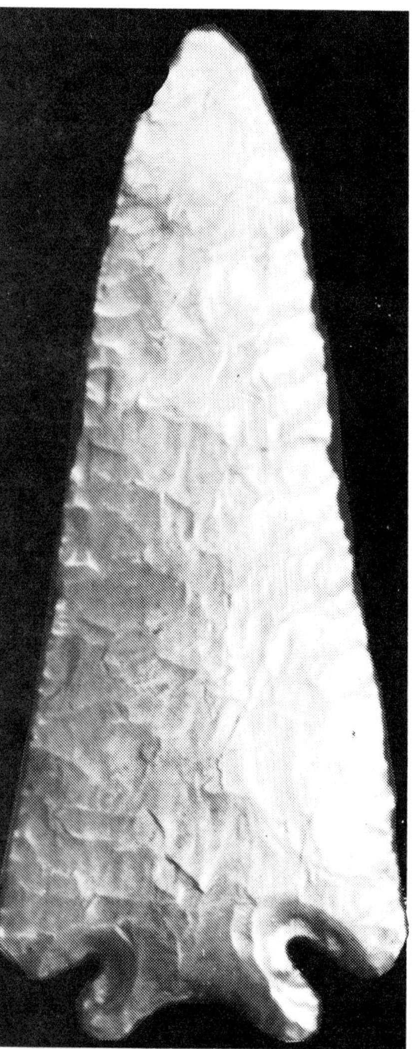

128-A - $1,500.00

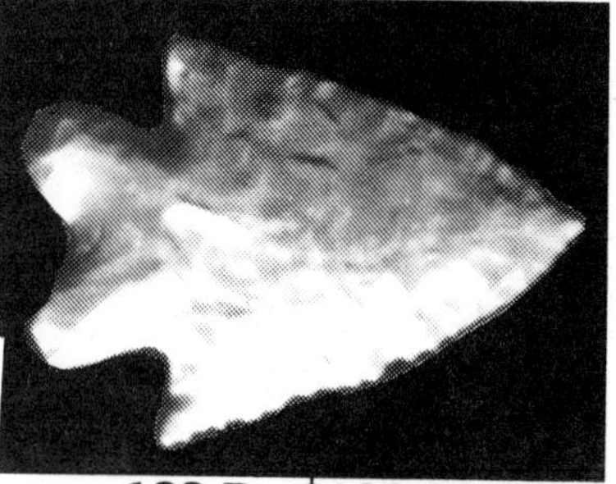

128-B - $100.00

Sublet Ferry

(Pictures 129-A thru 129-C)

Sublet Ferry is a small to medium sized point that is serrated and side notched. Average length is around 1-7/8 inches long and an average width of 3/4 inch.

Cross-section is biconvex. Blade edges tend to be excurvate but on some examples may be parallel angular, with an acute distal end. Shoulders are tapered fairly narrow. The base is side notched near basal edge. Shoulders and expanded stem tend to be nearly same width. Type has straight and thinned basal edge.

Blade and stem were shaped by removal of shallow to deep random flakes. Blade edge was serrated by removal of long, narrow, well spaced flakes. Stem was formed by removal of a large deep flake from both sides of each face just above base, which was sometimes retouched. Basal edge was then thinned by removal of broad thin flakes.

The type was named for points from sites around Sublet Ferry on the Tennessee River. Cultural material found on sites with this type point to an early Woodland or as old as early Archaic. Hard evidence on age of type is lacking, but most likely is around early Woodland period culture.

129-A - $45.00 **129-B - $50.00** **129-C - $15.00**

Suwannee

(Picture 130-A)

A large auriculate point. Hafting area is not as constricted as in the Simpson. Blade edges are excurvate. Hafting area is auriculate. Overall workmanship is not as good as the Simpson and has some resemblance to the Quad group.

130-A - $1,500.00
3 Inch example same quality ... $750.00
2 Inch example ... $200.00

Swan Lake

(Pictures 131-A thru 131-F)

Swan Lake is a small, somewhat thick, shallow side notched point. Average length is from 1-1/8 inches to 1-1/2 inches long and 1/2 inch to 3/4 inch in width.

Cross-section is biconvex though on some examples, a rough median ridge may be present. Blade edges tend to be straight, but may be incurvate or excurvate. Distal end is acuminate. Shoulders are narrow and usually tapered but may be expanded. Shallow side notches placed just above basal edge leave an expanded base stem for hafting. Side edges of stem are incurvate with an excurvate or straight basal edge. Points of this type often retain some of the nodular rind, on the basal edge, which are often left unfinished and may occasionally be ground.

Shaping of the blade and stem faces employed short, random flaking with short, deeper flaking used to retouch edges of blade and stem. Notches were formed by removal of several, deep, flakes just above basal edge on the side of the blade.

The type was named from sites in the Swan Lake area of Limestone County, Alabama where examples were first noted. Swan Lake may be related to point types such as Trinity and Halifax.

(continued on next page)

The Lamoka type, found mainly in New York is believed to be an early variation on this type. The Lamoka side notched has been carbon dated to between 4,500 to 5,500 years old. This relationship as well as the context of numerous excavated examples in Alabama, suggests an origin in the late Archaic period and increasing occurrence in the Woodland period.

131-A	131-B	131-C	131-D	131-E	131-F
$5.00	$10.00	$5.00	$15.00	$15.00	$10.00

Table Rock

(Picture 132-A)

A medium to large stemmed point. Stem edges are square to slightly expanded. Most examples are ground on basal edges. Shoulders are pronounced. Blade edges are straight to convex. Some collectors call these points Bottle Necks.

132-A - $350.00

Tallahassee

(Pictures 133-A thru 133-D)

The type is Dalton-like in appearance but is placed in the Gulf formational period, often with Fiber Tempered Pottery. The blade edges are serrated. Hafting area is usually expanding. Basal edge is incurvate and often ground.

133-A	133-B	133-C	133-D
$400.00	$200.00	$125.00	$55.00

Thebes

(Pictures 134-A thru 134-E)

A medium to large point type. Blade edges are straight to slightly recurvate or may be excurvate. Stem is corner to nearly side notched. Notches are usually deep. Basal edge may be incurvate. Workmanship is usually very good. Occasionally the blade edges are alternately beveled. Age - early Archaic.

(points on next page)

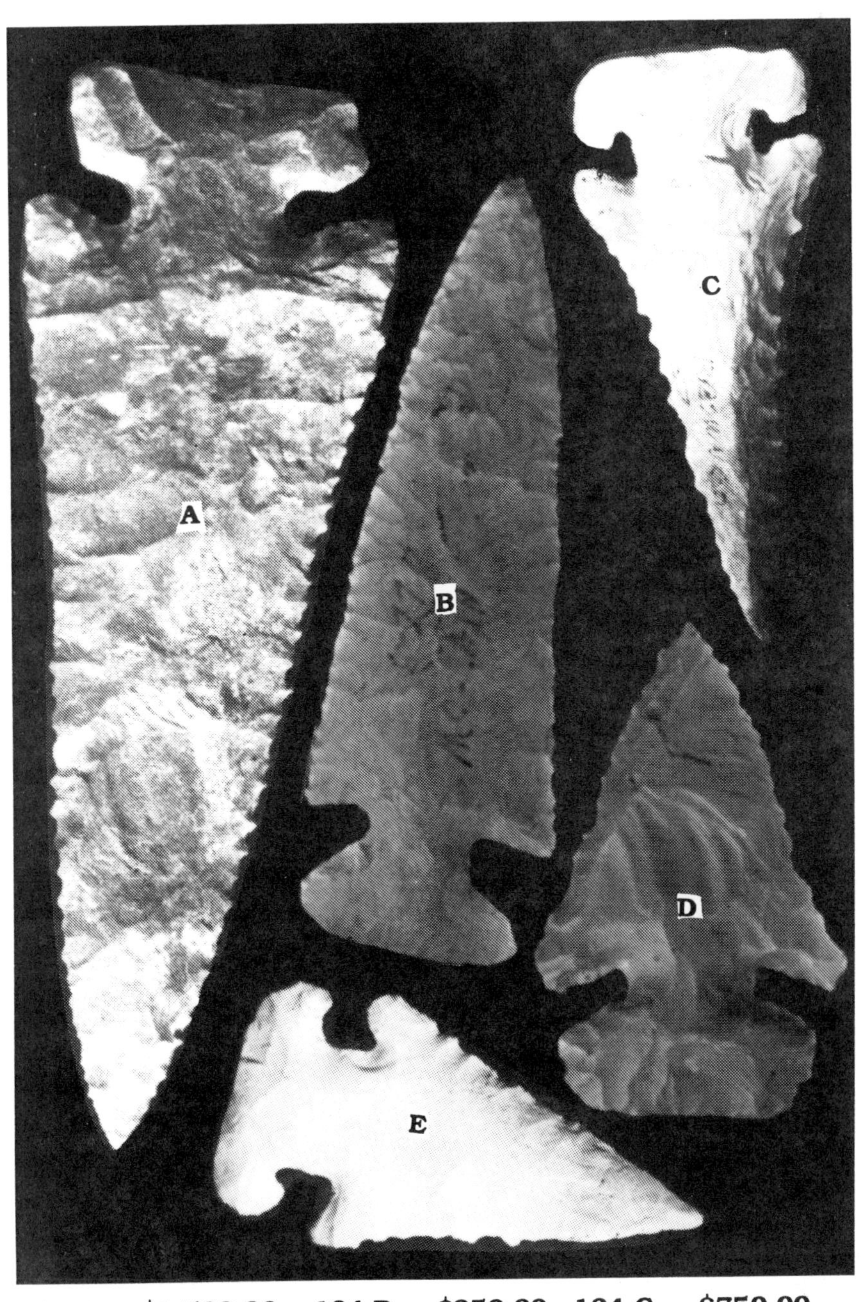

134-A - $1,500.00 134-B - $250.00 134-C - $750.00
134-D - $150.00
134-E - $350.00

Thonotosassa

(Pictures 135-A thru 135-B)

Often crudely made with weak or almost non-existent shoulders. A medium to large size point. Some examples have needle-like tips.

135-A - $50.00 (Broken)
135-B - $150.00

Turkey Tail

(Pictures 136-A thru 136-F)

Turkey Tail is a medium to large point with excurvate blade, side notches, and pointed on both ends. Average length is from 2 inches to 8 inches long, averaging around 5-1/2 inches. Average width is around 1-3/4 inches.

(continued on next page)

Cross-section is flattened. Blade edges are excurvate with acute or broad points on distal end, as well as the hafting end. Shoulders may be either horizontal or tapered. The pointed basal stem is thinned. There have been a very small number found with double notches.

Blade faces are shaped by well executed, broad, shallow, random flaking. Examples usually have short deep secondary flaking along all edges. From one to several flakes were removed to form the shallow-sided notches used for hafting.

The type was named for the similarity of the hafting area to the shape of a turkeys tail. The type as it is found in Alabama, appears to have been ceremonial in nature. It is usually associated with shell-mound burials, being found commonly in caches from a few points to 40 or more. The origin of this type appears to be late Archaic to early Woodland period, about 4,000 to 2,500 years.

136-A - $500.00
136-B - $750.00
136-C - $450.00

136-D - 550.00 136-E - $1,500.00

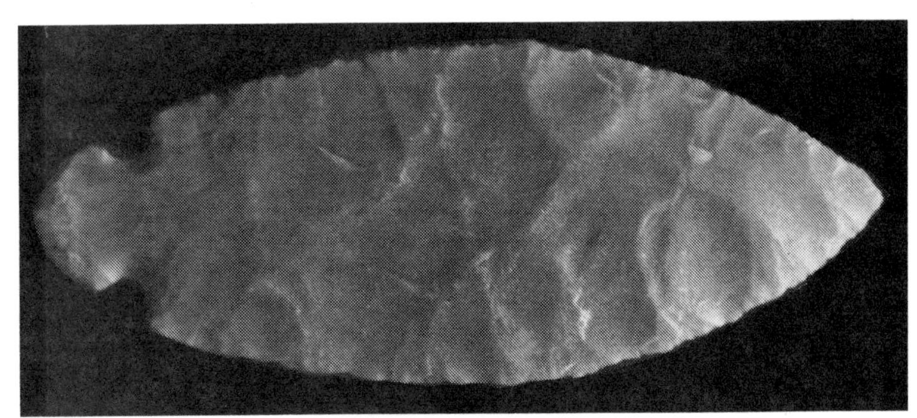

136-F - $750.00

Union Side Notched

(Pictures 137-A thru 137-B)

A medium sized auriculate point. Probably it is part of the Suwannee cluster. Side notches are shallow. The base is ground in most examples. Found in southern Georgia, Alabama, and Florida.

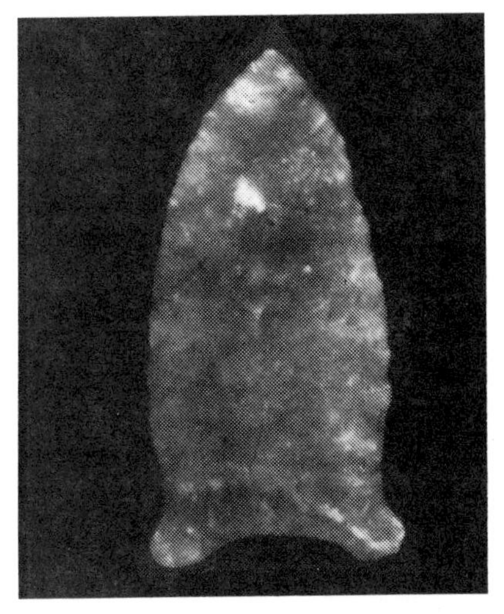

137-A - $250.00

137-B - $175.00

Wade

(Pictures 138-A thru 138-F)

Wade is a medium sized barbed point, with straight stem. Average length is from 1-1/2 inches to 2-3/4 inches long and from 1-1/8 inches to 1-1/2 inches in width at shoulders.

Cross-section is biconvex but may be flattened. Blade edges may be excurvate or straight. The barbs of the shoulders are at times as long as the stem. Examples are occasionally found with one long barb and one short barb. Distal end is acute. The stem usually has straight side edges with a straight or slightly incurvate basal edge. Stem is thinned and often lightly ground.

Flaking used to shape the blade and stem was deep to shallow, and random. Short, somewhat deep flaking was then applied to edges of the blade and stem. The notches were started by removing one large deep flake from each corner of each face. These notches were then finely retouched with small, fairly steep flakes being removed to complete the shaping of barbs.

The type gets its name from points found on several sites around Wade Landing on the Tennessee River. At these sites in Limestone County, Alabama, the type was first recognized in 1960. Wade has been found with a pre-ceramic culture burial as well as numerous Archaic and Woodland associated artifacts on sites in Alabama and Tennessee. From this and taking note of workmanship and physical features, there appears to be a late Archaic to middle Woodland origin from around 4,500 to as late as 3,500 years before present.

138-A - $150.00 138-B - $100.00 138-C - $150.00

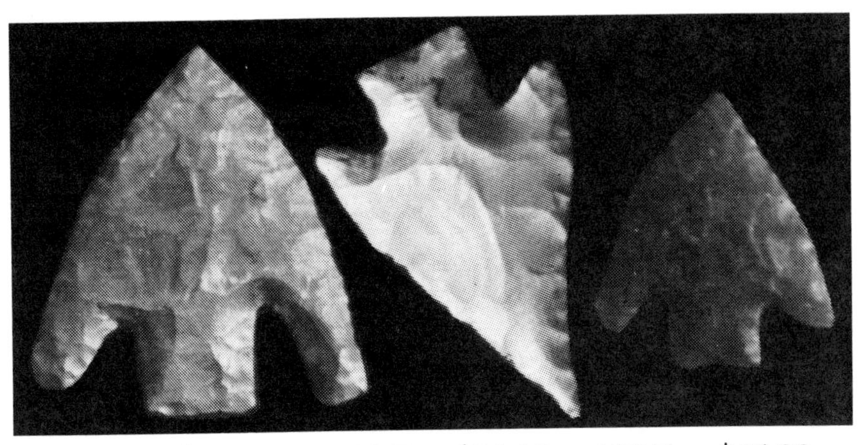

138-D - $150.00 **138-E - $50.00** **138-F - $45.00**

Washington

(Pictures 139-A thru 139-C)

Washington is a small, side notched point with serrated blade. Average length is from 5/8 inch to about 1 inch long and 3/8 inch to 1/2 inch in width at shoulders.

Cross-section is biconvex. Blade is excurvate or straight and most examples will be serrated. Shoulders are always inversely tapered. The distal end may range from acute to sharply acute or broad. The type is hafted by narrow, broad, side notches taken from each side of blade near basal edge, forming an expanded stem. The stems base may be excurvate or straight and in some points, thinned.

Shallow, random flaking shaped the blade and hafting area faces. Broad deep flakes were then removed to form serrations on blade edges. The notches are flaked by removal of one wide deep flake on both sides of both faces, just above the base. Some retouch of fine flaking occurs on edges, particularly near the distal end.

The type is named from sites in Washington County, Alabama, where type was first noted. In this area, type was densely distributed occurring on surface sites in association with Woodland assemblages. The material most often encountered is white quartzite and generally shows good workmanship, especially considering the difficulty with which quartzite is normally worked.

(points on following page)

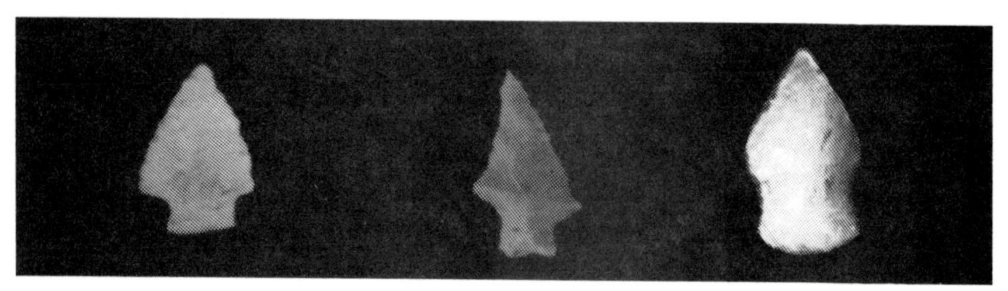

139-A - $5.00 139-B - $5.00 139-C - $5.00

Washita

(Picture 140-A)

Washita is a small triangular point with side notches. Average between 3/4 inch and 1 inch in length and 1/2 to 3/4 inch in width.

Cross-section is flattened. Blade edges may either be straight or excurvate with an acute distal end. Between 1/3 and 1/4 the distance from base to distal end, there are side notches for hafting. Basal edge will be straight or slightly incurvate. Basal edge is thinned.

Relatively broad shallow random flakes were removed to shape the blade and hafting area. A minimum amount of retouch was applied to finish the edges. Short deep flakes were removed from each side of each face to form notches in sides of blade.

This type was named after points found to be associated with the Washita River Focus in Oklahoma. This type is very similar to the Harrel point, but lacks the notch in the basal edge. The type is distributed in Oklahoma, segments of the Great Plains, Mississippi Valley and scattered in the southwest. Only a few finds are made of this type outside these areas. The association with the Harrel point, pottery and agricultural sites where it is found indicates an early Mississippian period culture.

140-A - $10.00

Wheeler

Excurvate, Recurvate, Triangular and Incurvate

(Pictures 141-A thru 141-S)

Wheeler excurvate is a small to medium sized point with deep incurvate base. Average length is from 1 inch to 2-1/2 inch with a median length about 1-3/4 inches and averages around 3/4 inch in width.

The cross-section is most often biconvex but may be plano-convex. Distal end is acute. Hafting area sides show little or no change where blade joins haft. Grinding may be only around hafting areas or at times may be all the way to distal end of blade edge. The base is parallel pointed and deeply incurvate on basal edge. The pointed auricles may be slightly contracted at points of basal edge. Some examples may be fluted.

Well controlled shallow, random flaking was used to shape faces of blade and hafting area. Blade edges were further shaped by removal of fairly broad but regular flaking along edges, then a fine retouch refined blade edges. Basal edge has large flakes removed to form cavity in base, and was then finished by removal of short steep flakes, leaving basal edge steeply beveled.

The type gets its name from the Wheeler Basin of the Tennessee River where points of type were recognized. The type appears on pre-shell-mound sites in north Alabama in association with Paleo and transitional Paleo period. All members of the Wheeler family are quite rare.

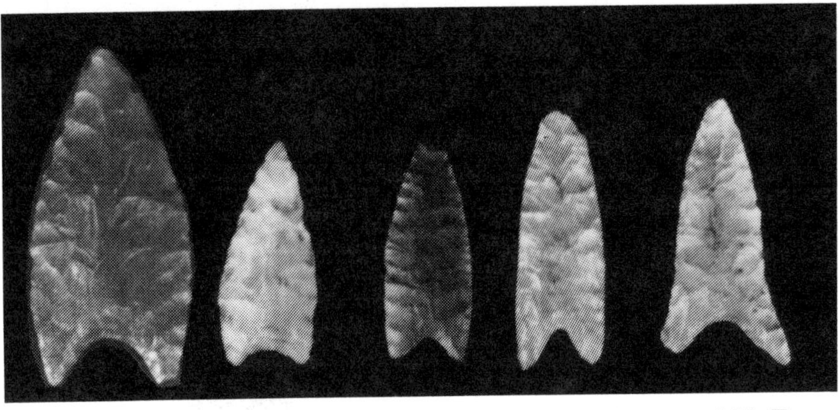

141-A	**141-B**	**141-C**	**141-D**	**141-E**
Excurvate	*Excurvate*	*Excurvate*	*Excurvate*	*Incurvate*
$200.00	**$75.00**	**$50.00**	**$50.00**	**$75.00**

141-F	141-G	141-I	141-J	141-K
Incurvate	*Recurvate*	*Recurvate*	*Recurvate*	*Recurvate*
$200.00	**$125.00**	**$300.00**	**$100.00**	**$250.00**

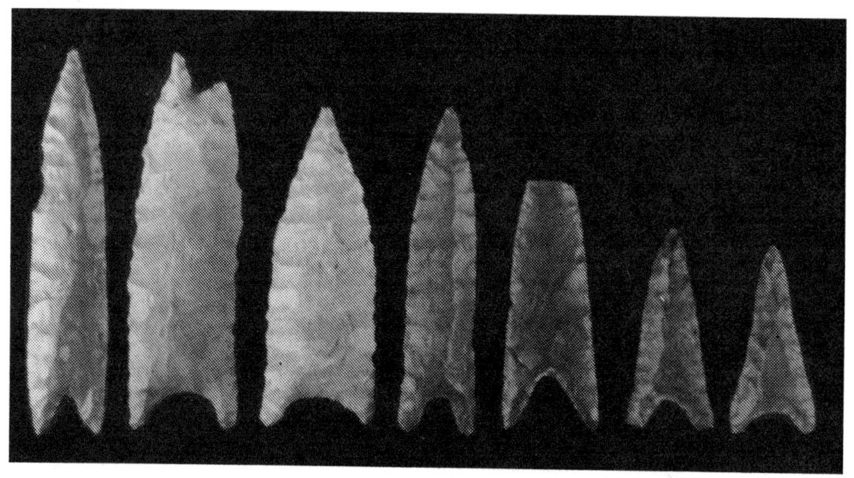

141-M	141-N	141-O	141-P	141-Q	141-R	141-S
$300.	**$400.**	**300.**	**$300.**	**$75.**	**$75.**	**$75.**

All of these are Triangular

Points 141-E and 141-F have been called Expanded Base Wheeler by some collectors. However, the author feels the term Incurvate should be used in order to stay within the theme set by Cambron.

141-H - $150.00 - *Recurvate*
141-L - $1,500.00 - *Triangular*

White Springs

Wacissa In Florida

(Pictures 142-A thru 142-G)

White Springs is a medium sized broad stemmed point with excurvate blade edges. Average length is between 1-5/8 inches and 2-1/2 inches long and from 1-1/8 inches to 1-1/2 inches in width.

Cross-section is biconvex or flattened. Blade edges are slightly excurvate or, rarely, slightly incurvate. Distal end is acute. Shoulders are narrow and horizontal. Hafting area has a broad stem with straight sides and straight or slightly excurvate basal edge. Base is thinned and may be ground. Blade was shaped by shallow, random flaking and may sometimes exhibit transverse oblique flaking on faces. Edges of the blade and stem were then retouched with short pressure flaking.

The type is named for finds on sites around White Springs area of Limestone County, Alabama. The heaviest distribution of type is along the Tennessee River Valley in north Alabama. This type is found on predominately early Archaic sites with such types as Eva, Morrow Mountain, Benton and Buzzard Roost points. Some believe the White Springs to be older than the Benton and Buzzard Roost points. Occurrence in the early Archaic around 6,000 to 7,000 years ago seems to be indicated.

(points on following page)

142-A - $150.00 142-B - $75.00 142-C - $100.00

142-D-$45.00 142-E-$75.00 142-F-$45.00 142-G-$25.00

Yadkin

(Pictures 143-A thru 143-B)

A triangular shaped point. Blade edges may be straight or recurvate. Sometimes has serrated edges. Found on Woodland sites from South Carolina to southern Georgia.

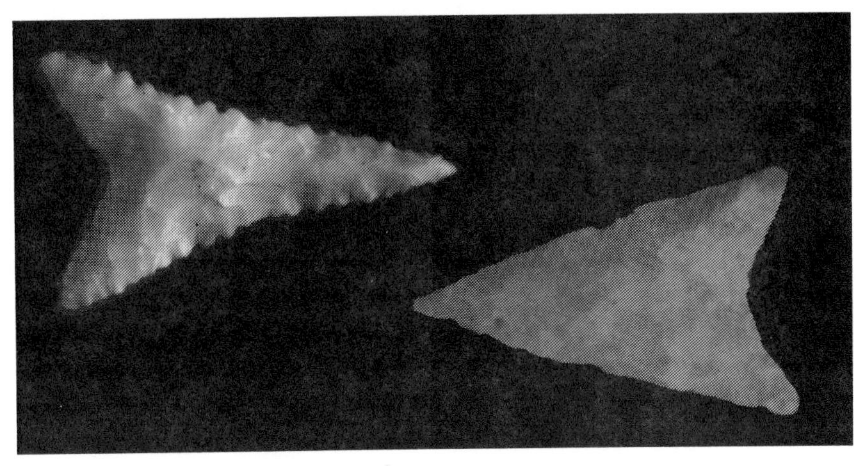

143-A - $125.00 **143-B - $35.00**

Louisiana Arrow Point Types

Small Arrow points from Texas, Oklahoma, Arkansas, Louisiana.

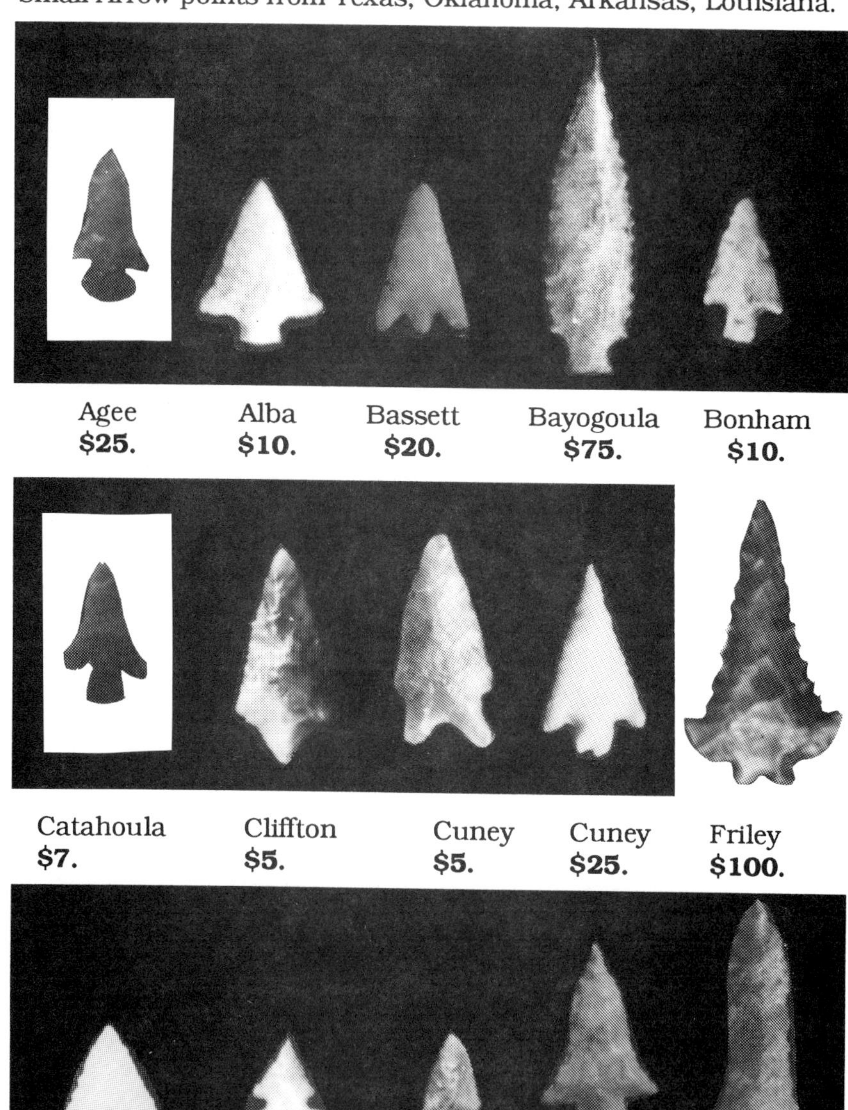

Agee **$25.** Alba **$10.** Bassett **$20.** Bayogoula **$75.** Bonham **$10.**

Catahoula **$7.** Cliffton **$5.** Cuney **$5.** Cuney **$25.** Friley **$100.**

Garza **$25.** Harrell **$25.** Haskell **$10.** Hayes **$45.** Homan **$40.**

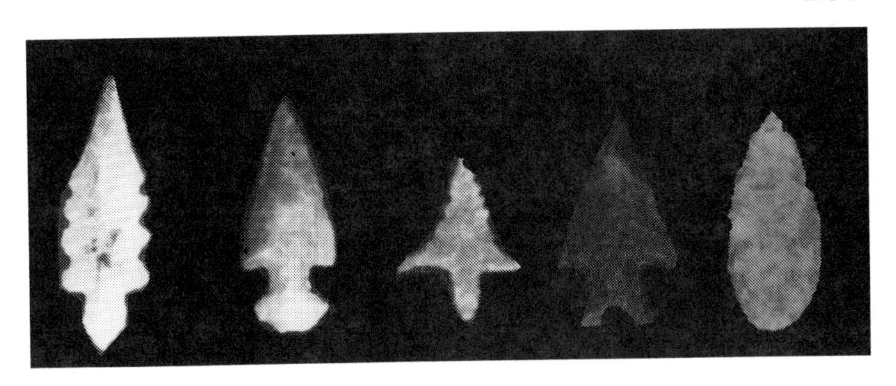

Howard
$100.

Keota
$45.

Livermore
$7.

Morris
$25

Nodena
$10.

Perdiz
$20.

Reed
$10.

Sallisaw
$35.

Scallorn
$10.

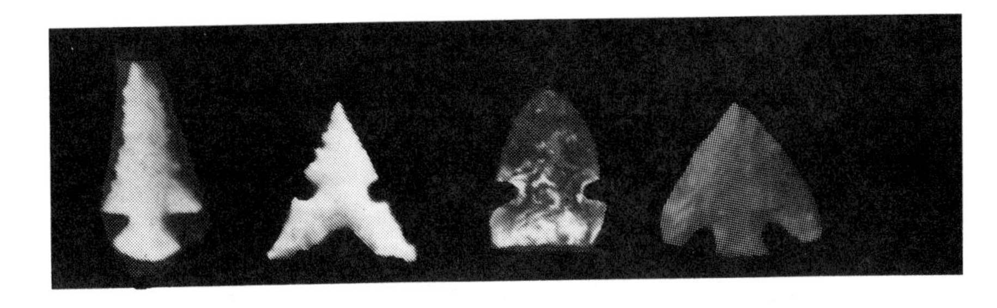

Sequoyah
$10.

Toyah
$40.

Washita
$10.

Deadman
$25.

Texas Point Types

Abasolo-1 Almagre-2 Angostura-3

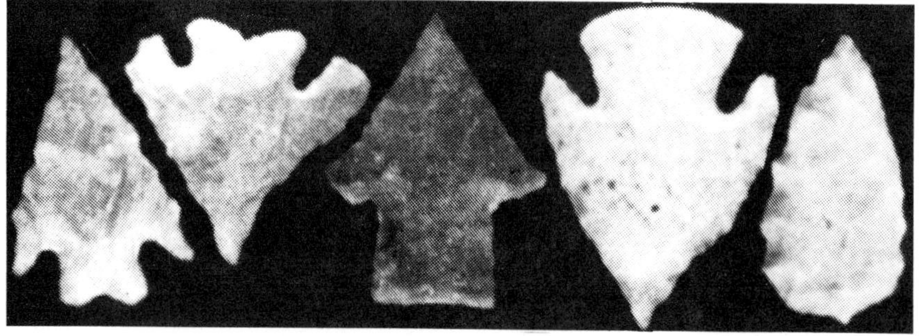

Baker-4 Bandy-5 Bulverde-6 Castroville-7 Catan-8

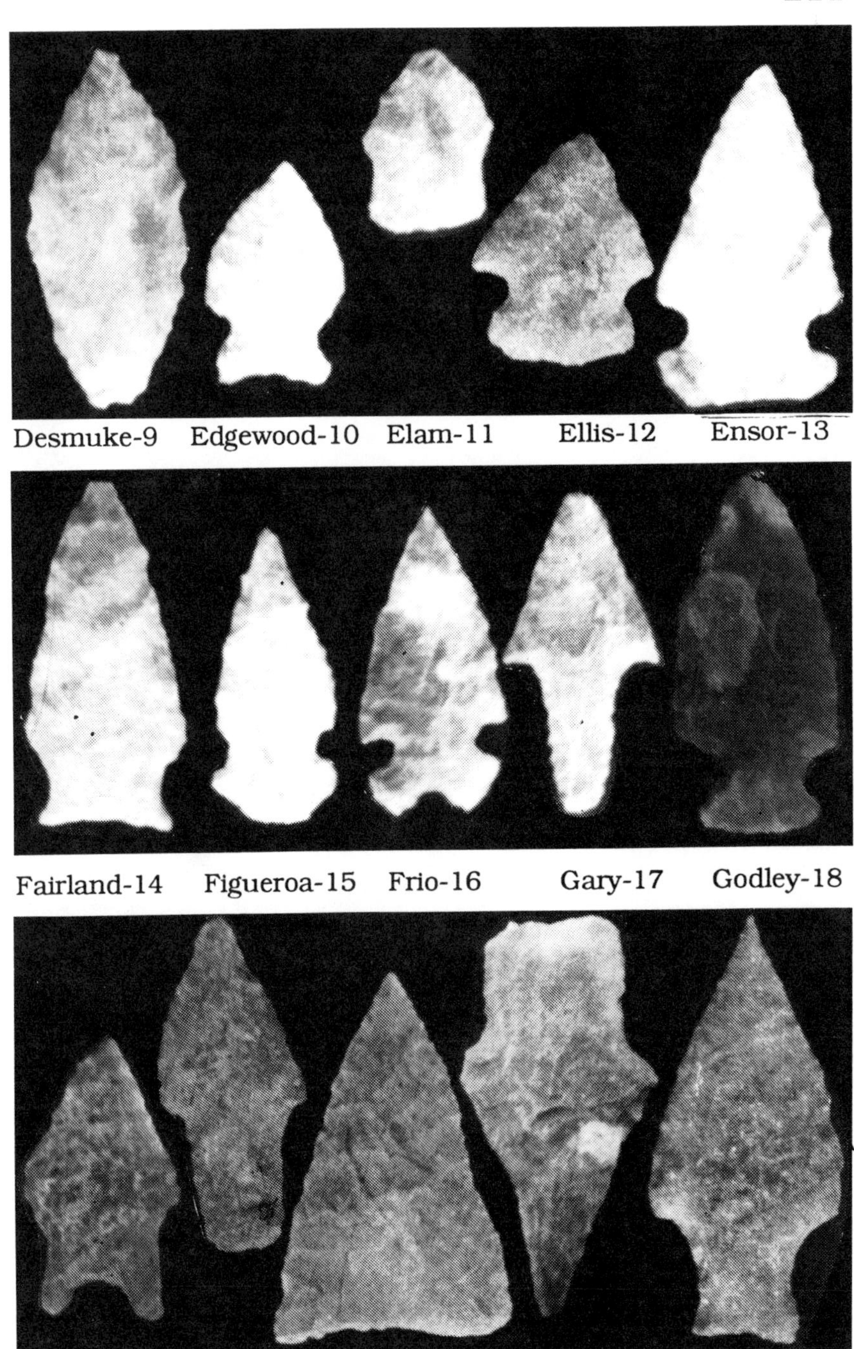

Desmuke-9 Edgewood-10 Elam-11 Ellis-12 Ensor-13

Fairland-14 Figueroa-15 Frio-16 Gary-17 Godley-18

Gower-19 Kent-20 Kinney-21 La Jita-22 Lange-23

Langtry-24 Lerma-25 Marcos-26 Marshall-27

Martindale-28 Matamoros-29 Meserve-30 Montell-31Morhiss-32

Nolan-33 Paisano-34 Palmillas-35 Pandale-36 Pandora-37

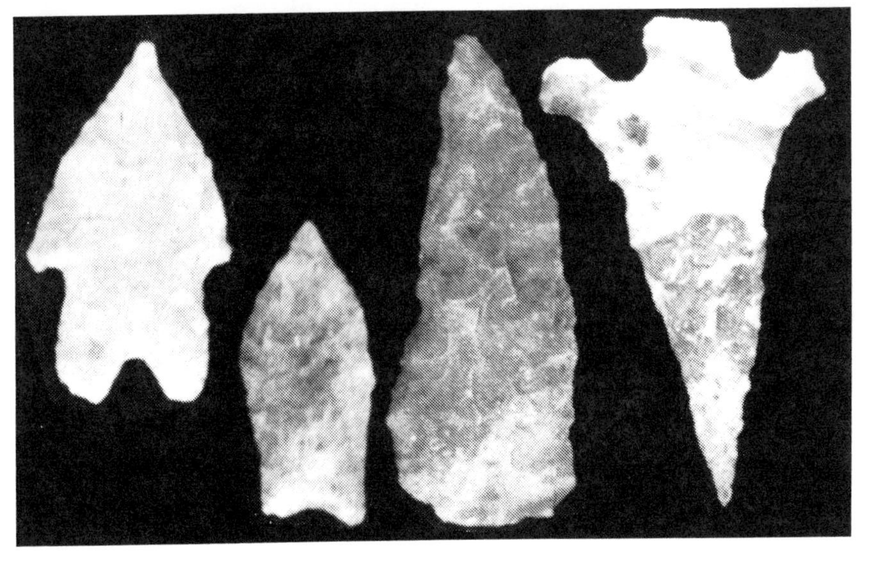

Pedernales-38 Plainview-39 Refugio-40 Shumla-41

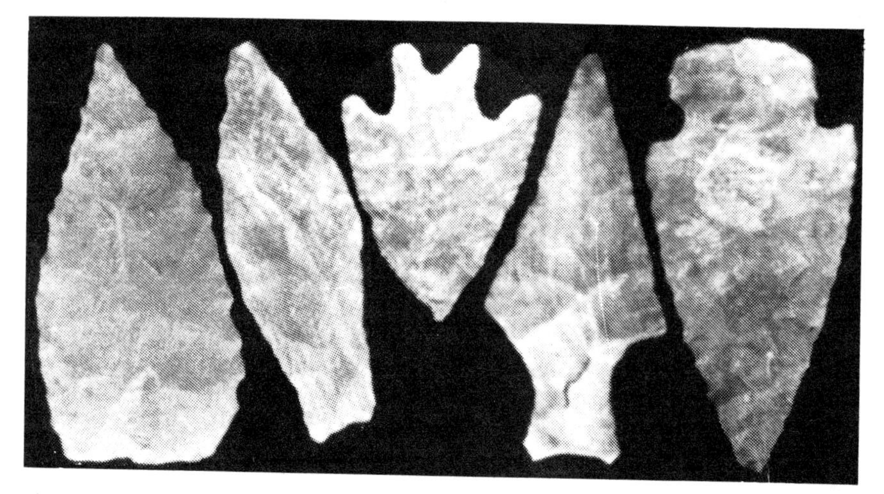

Tortugas-42 Travis-43 Uvalde-44 Val Verde-45 Williams-46

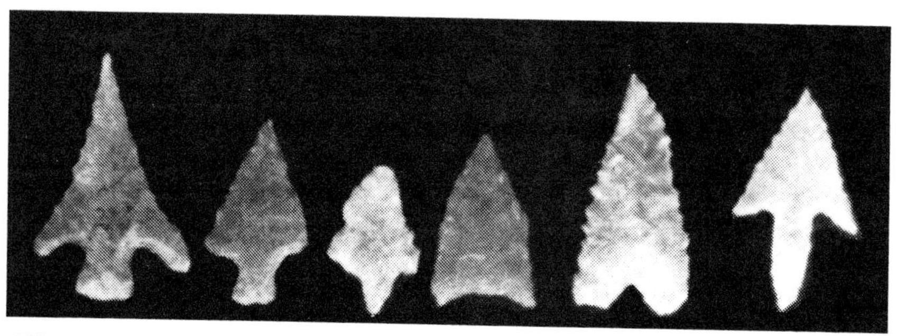

Alba Bonham Cliffton Fresno Garza Perdiz

BONE

The pointed, splintered end of a piece of bone was probably one of mans first tools. Without any other processing, it could be used as a needle, spear point, or dagger. It is the authors belief that the flute in Paleo points was an attempt to duplicate the marrow channels which would have been found on early points made of bone. The related materials such as antler, horn, turtle shell, ivory and the teeth and claws of animals, were universally used by all the Indians. These materials were used not only for tools and ornaments, but also certain bones and skulls were used by medicine men for spells and rituals.

Stone implements such as knives, scrapers, saws, gravers, grinding stones and drills were used to shape bone and similar materials.

Paleo man may even have used the bones of large animals, such as mammoths, for supports in their houses. We know stone age man in Europe used the bones of mammoths for this purpose.

Not only did the Indians use the bones of animals, they also used the bones of their captives and enemies who were killed in battle.

It is not uncommon to use a cup made from a human skull to eat and drink out of. It was thought that one could acquire the knowledge and wisdom of an admired enemy by doing so.

The Hopewellion culture made further use of human bone by carving intricate designs on hairpins carved from human leg and arm bones.

The practice of using human bone was first noted during the Archaic period and continues into the Mississippian period. Archaic people were the first that we know of who took trophy skulls. These are sometimes found in burial association. A few of them still have traces of painted designs. Archaic people also made gorgets and pendants from human skulls. These gorgets often have geometric designs engraved on the convex surface.

The practice of taking trophy skulls or heads continues into the Mississippian period. The head pot is the result of molding a representation of the features of the skull while still in the flesh. Most head pots are death effigys with swollen features and a fold of skin on the forehead which was pierced with a bone needle. This fold of skin was probably used to suspend the head from a trophy pole.

7

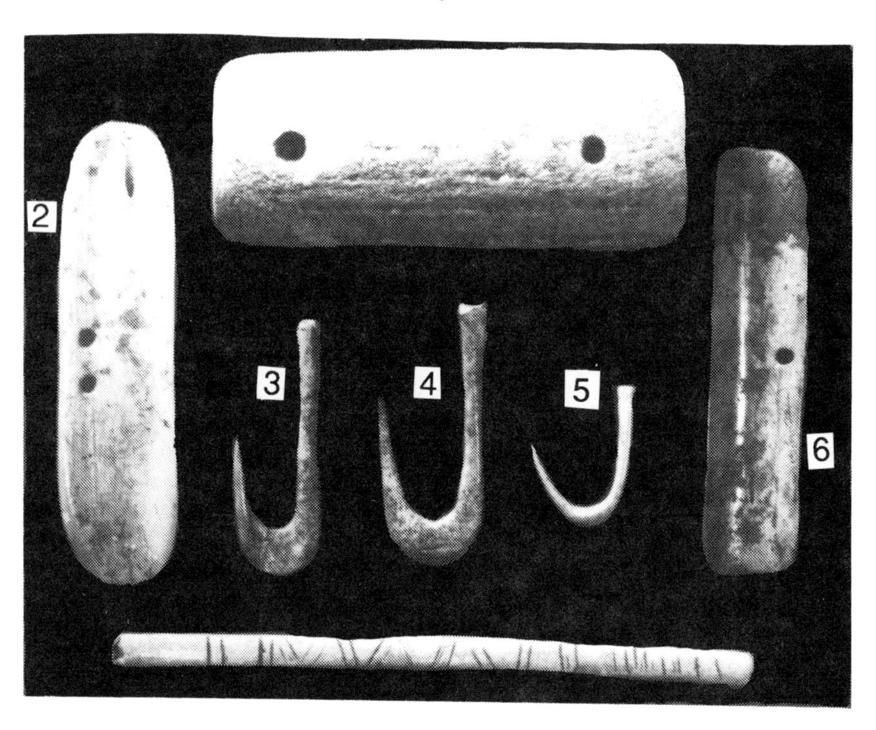

1

1.	**Gorget Deer Antler**	**$275.00**
2.	**Gorget**	**$250.00**
3.	**Fish Hook**	**$150.00**
4.	**Fish Hook**	**$150.00**
5.	**Fish Hook**	**$75.00**
6.	**Gorget**	**$200.00**
7.	**Engraved Pin**	**$300.00**

8. **Gorget 3-1/2 inch**
Made from human skull
Archaic - **$2,500.00**

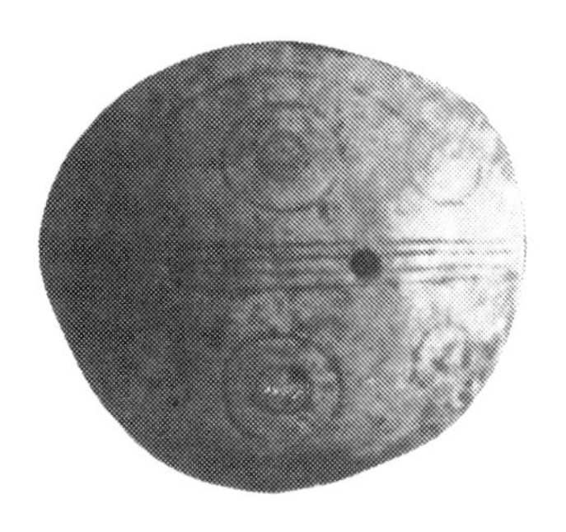

9. **Gorget 3 inch**
Made from human skull
Archaic - **$2,000.00**

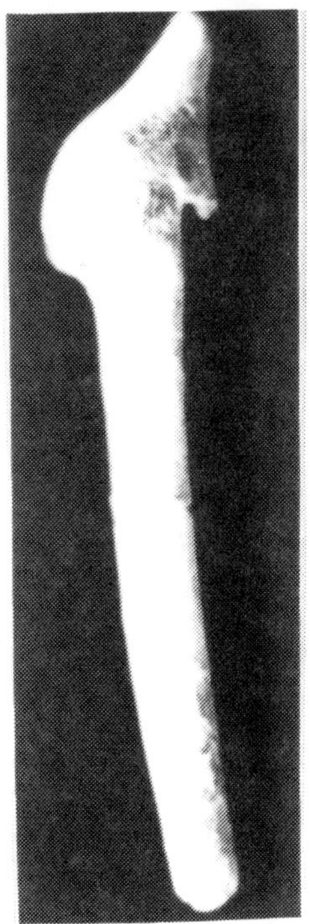

10. Atlatl Hook
Deer Antler, broken
$350.00

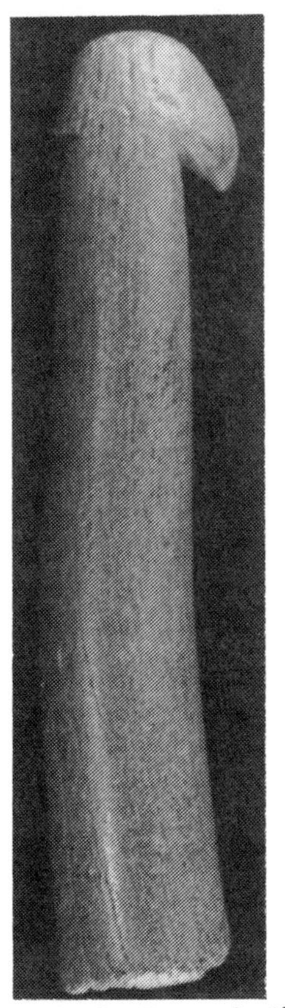

11. Atlatl - Deer Antler

$750.00

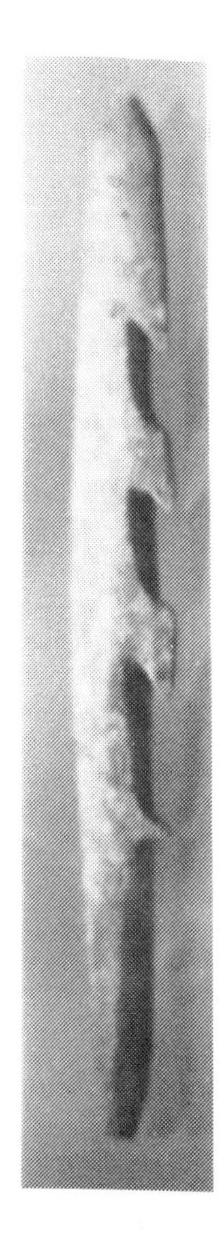

12. **Bone Harpoon**
$450.00

13. **Deer Antler Projectile**
$150.00

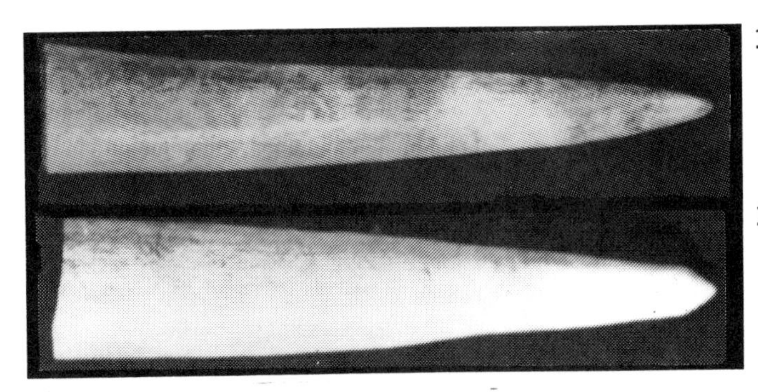

14. Deer Antler Projectile $125.00

15. Deer Antler Projectile $100.00

16 17 18

Bone Projectile - Paleo ... $350.00 to $500.00 each

19

20 21 22 23 24 (bottom) 25 26 27 28

19.	**Bone Projectile - Archaic**	**$150.00**
20.	**Atlatl Hook-Side View of #10**	
21.	**Atlatl Hook**	**$500.00**
22.	**Deer Antler Projectile**	**$125.00**
23.	**Bone Projectile**	**$75.00**
24.	**Atlatl Hook, engraved with cross hatching**	**$450.00**
25.	**Atlatl Hook, broken end**	**$300.00**
26.	**Turkey Bone Whistle, or turkey call**	**$250.00**
27.	**Flaked Bone Projectile, probably Paleo, very rare**	**$500.00**
28.	**Gorget, Archaic**	**$175.00**

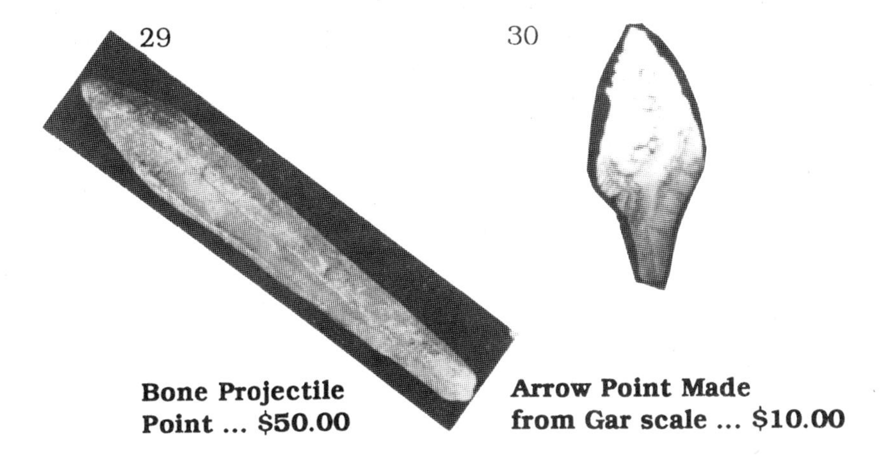

Bone Projectile Point ... $50.00

Arrow Point Made from Gar scale ... $10.00

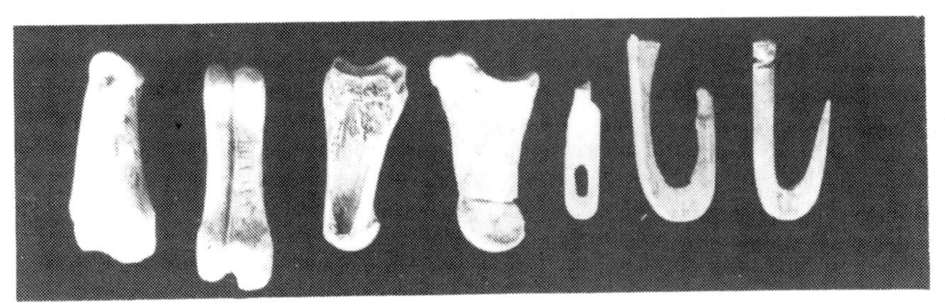

Steps in the process of making a fish hook from a deer toe bone.

31 32 33 34 35 36 37

31.	**Toe Bone**	**$10.00**
32.	**Toe bone incised and ready to split**	**$45.00**
33.	**Split toe bone**	**$75.00**
34.	**Split toe bone line cut to remove distal end**	**$75.00**
35.	**This is a section of deer leg bone some process works for both types of fish hooks**	**$100.00**
36.	**Blank nearly completed**	**$100.00**
37.	**Finished Hook**	**$150.00**

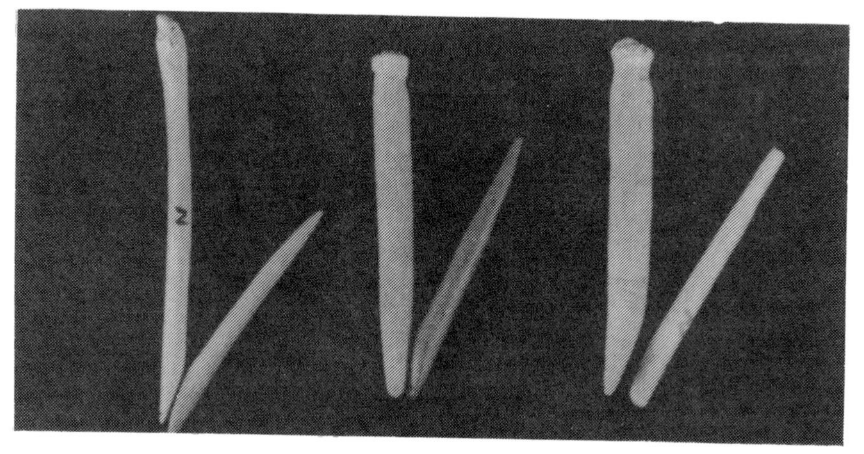

38. Two Piece Fish Hook $250.00
39. Two Piece Fish Hook $250.00
40. Two Piece Fish Hook $250.00

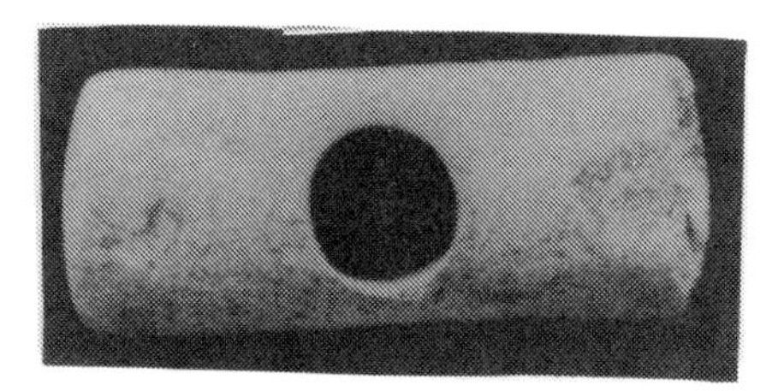

41. Antler Atlatl Weight, 3" $450.00

Drilled Pieces - Ankle Bone, Cannon Bone and Toe Bone For the Toss and Catch Game

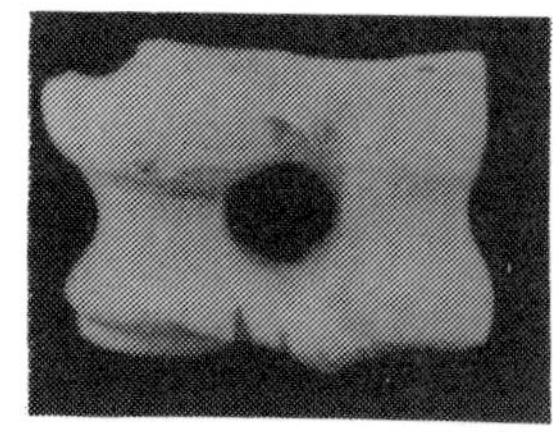

42. Ankle Bone

$75.00

43. Cannon Bone

$75.00

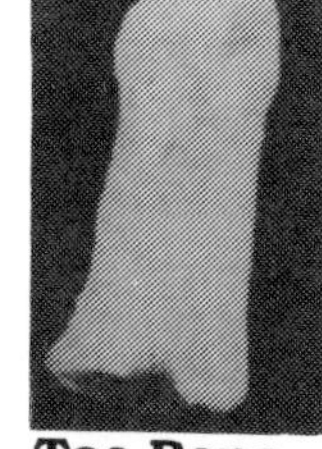

44. Toe Bone, drilled end to end

$75.00

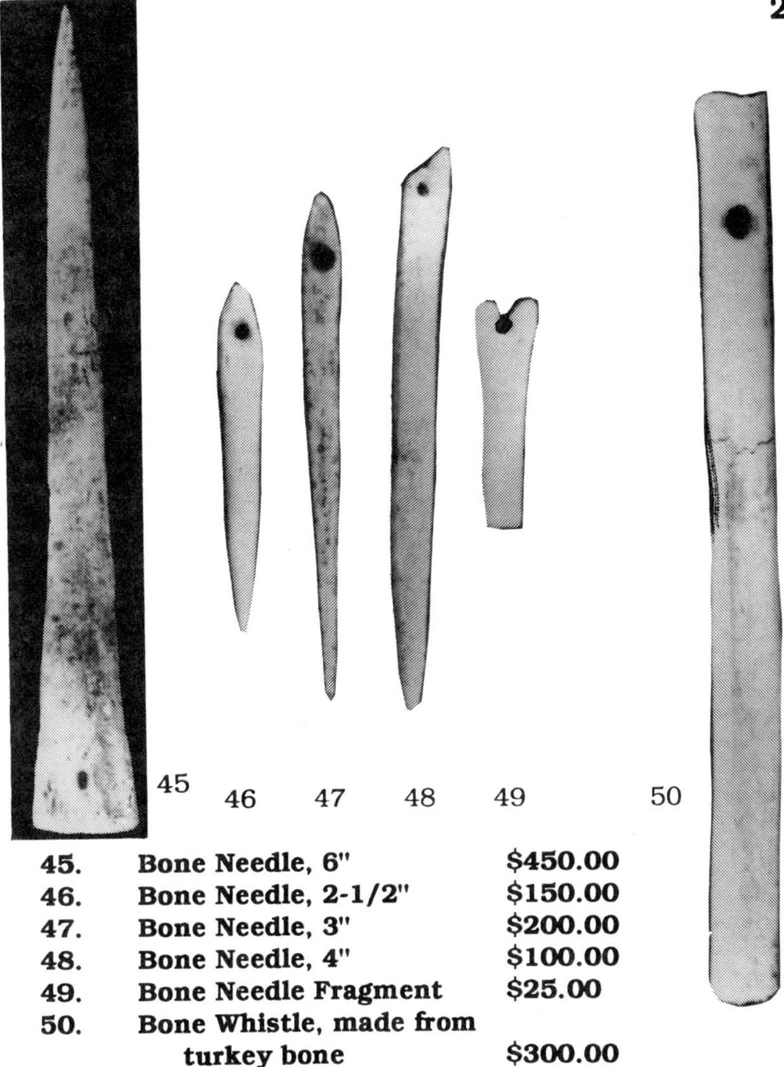

45.	Bone Needle, 6"	$450.00
46.	Bone Needle, 2-1/2"	$150.00
47.	Bone Needle, 3"	$200.00
48.	Bone Needle, 4"	$100.00
49.	Bone Needle Fragment	$25.00
50.	Bone Whistle, made from turkey bone	$300.00

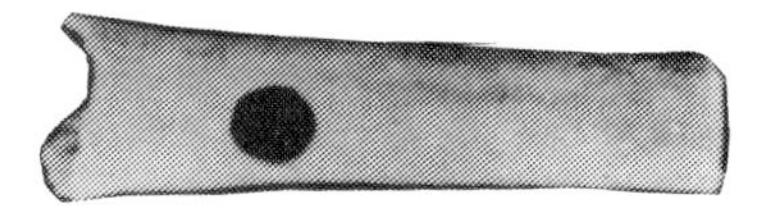

51. Bone Whistle

$125.00

52. Drilled Teeth
$15.00 to $150.00 depending on size and species

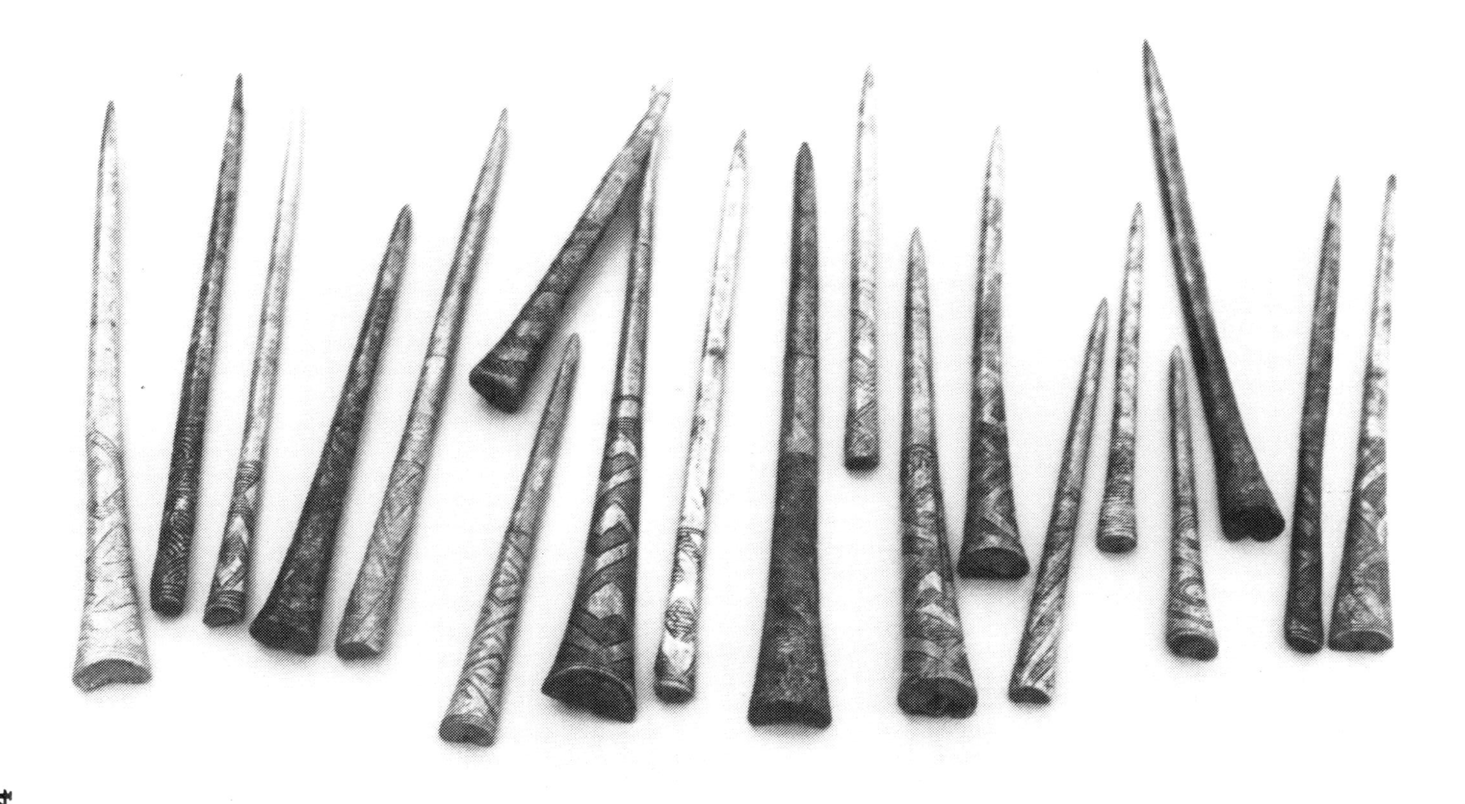

Archaic Bone Pins or Awls, Engraved ... $500.00 to $1,500.00

POTTERY

In North America, many of the more cultured tribes were skillful potters. There were two cultural centers that were more advanced than the rest. These were the Pueblo region of the Southwest and the Mound Builder culture of the Mississippi Valley and the Gulf states. Over most of the remainder of the United States and Canada, the art was limited to the construction of crude utensils or else was vertually unknown.

Pottery is among the last arts to be developed by primitive people. This is probably because pottery is not a very useful item to Nomadic people due to its fragility. It is only when a culture becomes sedentary that it becomes needed. Among more primitive people, boiling with hot stones in baskets and bark and wooden vessels, sometimes even bags made of skin, was practical.

As early potters began working clay into pots, they discovered that they had to add a tempering agent to the clay. Tempering is necessary to prevent the pot from shrinking and cracking while drying and firing.

Materials used to temper include fiber, sand, crushed stone, crushed potsherds and crushed shell. The earliest pottery was tempered with fiber and sand while Mississippian Indians used shell for tempering.

The production of pottery varied from tribe to tribe. Different cultures had pots of different shapes and designs were different. However, the basic steps in crafting a pot were usually the same. A small piece of clay was flattened into a concave disc. Then the walls were built up by adding strips of clay in a spiral fashion. This method is called coiling. Sometimes a basket or a piece of old pottery was used as a base to help mold and shape the bottom of the pot. This also allowed the pot to be turned as it was shaped and finished.

The Indians never developed the potters wheel or learned to glaze their pottery. In spite of this, Indian pottery ranks with the best artwork ever produced by any other culture.

Southeastern Pottery will bring 3 to 5 times more than Mississippi Valley Pottery. #1 thru #30.

1. Cord Stamped, Woodland 10" tall
$1,500.00

2. Jar on a bottle Effigy, Parkins
$750.00

3. Noded Saddle Pot
$900.00

4. Santa Rosa Swift Creek
$1,500.00

5. *Paddle Stamped with Lug Handles*
$1,200.00

6. *Irene Incised*

$1,900.00

8. *Dallas Incised*
$750.00

7. *Irene Incised*
$2,200.00

9. Fort Walton Incised
$2,000.00

10. Irene Incised
$950.00

11. Paddle Stamped
$1,100.00

12. Creek Incised
$1,200.00

13. Skull Effigy Bowl

$,1000.00

14. Strap Handle Jar
Incised Faces
$1,100.00

15. Noded Rim Jar
$650.00

16. Irene Incised
$1,200.00

17. Incised Jar
$1,500.00

18. Weeden Island Incised
$1,100.00

19. Irene Plain
$850.00

20. Dallas Frog Effigy Jar
$750.00

21. Dallas Strap Handle Jar
$450.00

22. Flying Rattlesnake Water Bottle
$3,300.00

23. Top View of Flying Rattlesnake Water Bottle

24. Beaker Engraved with Scalp Lock Pendants
$1,100.00

25. *Negative Painted Water Bottle*

$3,200.00

26. *Miniature Sandstone Tempered Jar*

$300.00

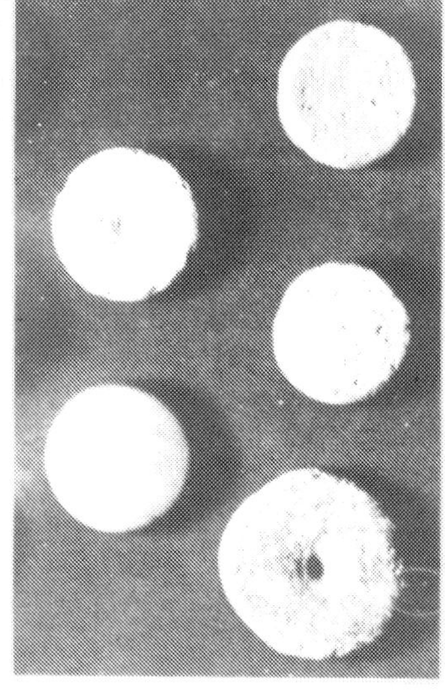

27. *Pottery Game Discs*

$10 to $25 each

28. *Alexander Incised and Stamped Shards, average value*

$20.00 ea.

30. *Large Saddle Pot*

$1,800.00

29. *Wattle, Fired Clay Lining of Indian Houses, size shown*
$25.00

MISSISSIPPI VALLEY POTTERY

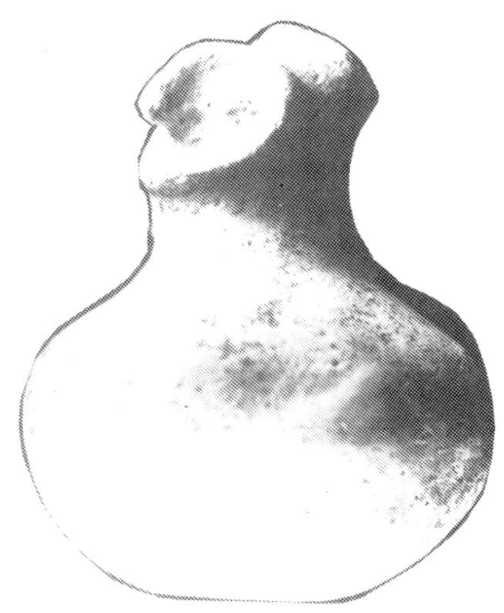

31. *Miniature Owl Effigy*
$375.00

32. *Rattlehead from large bowl*
$75.00

NOTE: Due to incorrect information the wrong prices were given for the following artifacts. The new prices are listed.

33. Walking Frog Teapot
Slight restoration
$2,200.00

34. Corn God
$750.00

35. Human Effigy Waterbottle
Slight restoration
$1,950.00

36. Double Headed Owl
Extremely Rare Piece
$1,550.00

37. *Human Effigy, Hunchback*
$1,200.00

38. *Corn God, Pressure Crack*
$750.00

39. *Marriage Jar, Rare*
$950.00

40. *Stirrup Bottle, Rare*
$925.00

41. *Bear Effigy Water Bottle, Miniature*
$300.00

42. *Human Rim Effigy, Incised Sides*
$825.00

43. *Who Bottle*
$275.00

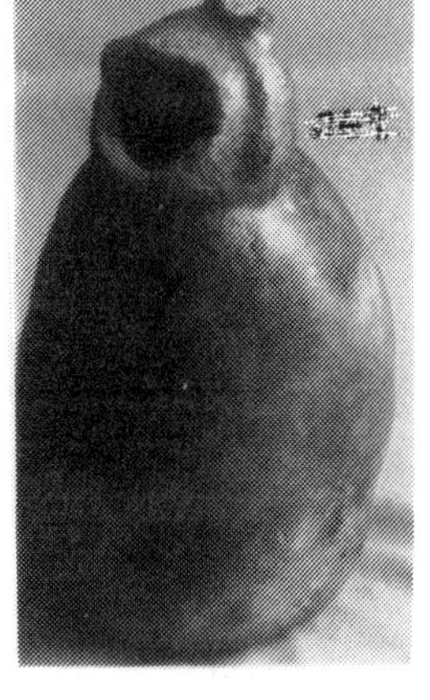

44. *Animal Head Rim Effigy*
$250.00 to $500.00

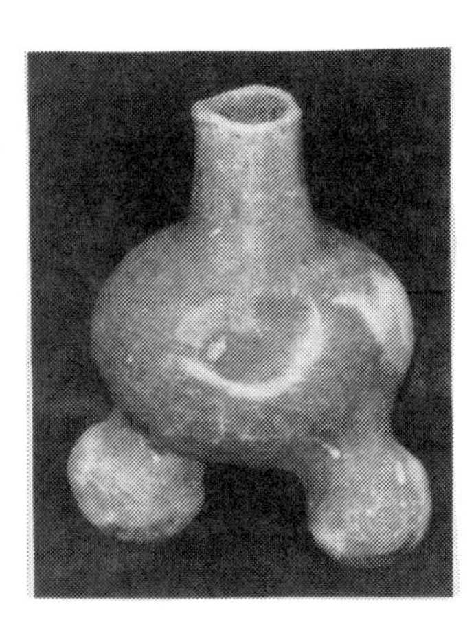

45. Tripod Bottle, 8"
$590.00

46. Tripod Bottle, 9"
$845.00

47. Parkins Strap Handle
$600.00

48. Rhodes Incised, rim damage
$650.00

49. Kent Incised with Parkins neck
$800.00

50. Painted Red on Buff Tripod
$860.00

51. Turtle
$600.00

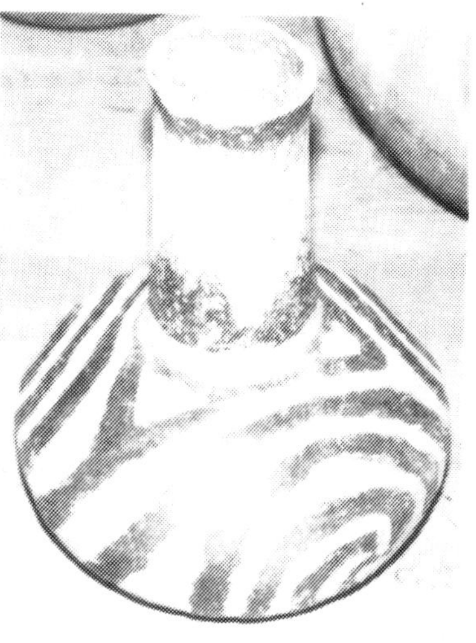

52. Painted Red on White
$5,000.00

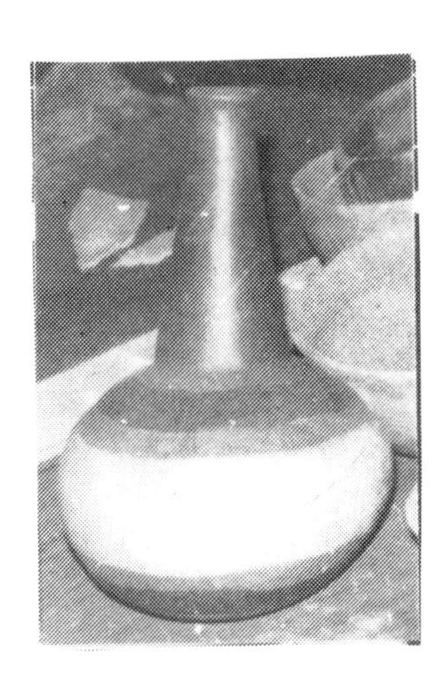

53. *Painted, Polychrome Red, Buff, White*
$7,500.00

54. *Strap Handle Jars*

$200 to $250 each

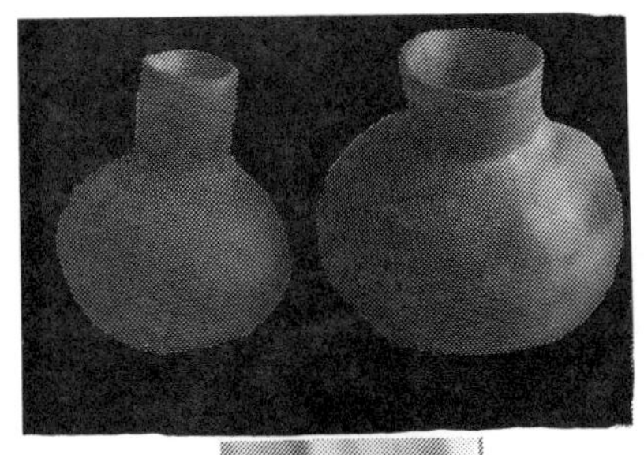

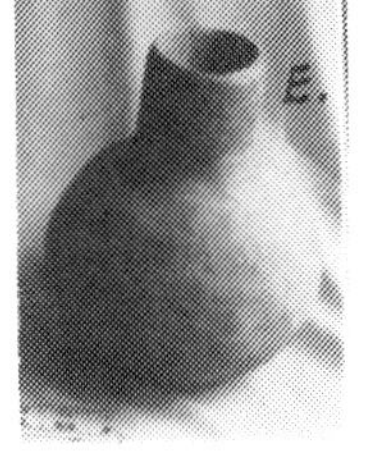

55. *Average Water Bottles*
$250 to $350 each

56. *Red and Buff Teapot*
$3,500.00

57. **a. $175.00**
b. $150.00
c. $100.00

58. Quawpaw, Polychrome Red, Buff and White
$3,000.00

59. Caddo Applique

$600.00

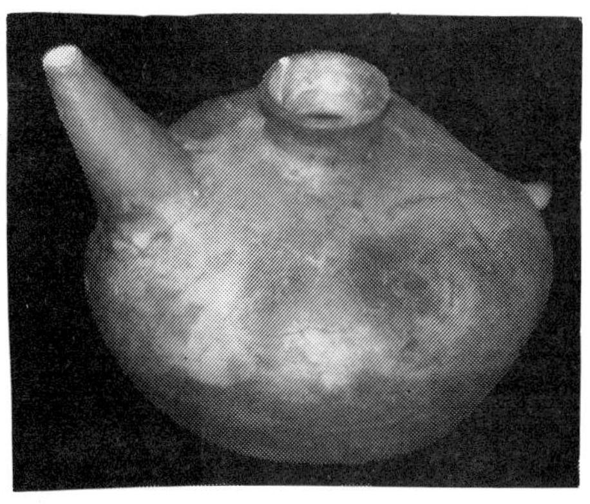

60. Teapot, broken and reglued
$800.00

61. Taylor Incised, broken
$275.00 (Intact, $900.00)

62. Haley Incised
$2,700.00

63. Anasazi, Tularosa Olla
$1,500.00

64. Anasazi, Lino Olla
$750.00

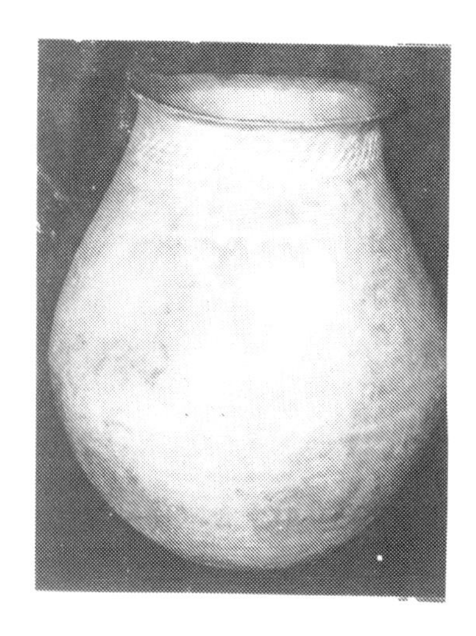

65. Anasazi, Lino Olla, large size
$5,000.00

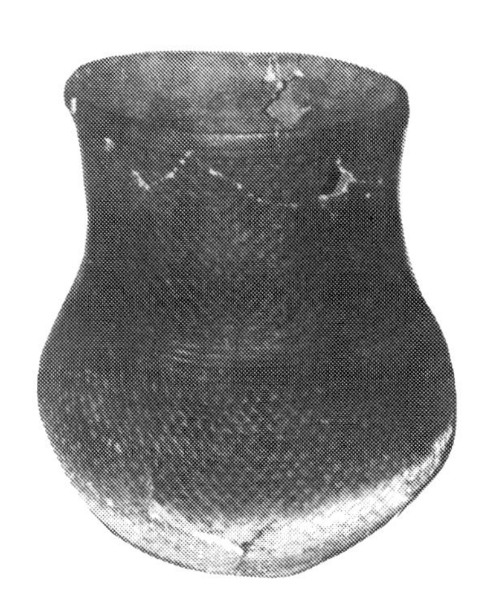

66. Tularosa, Olla 16x13 Damaged
$450.00

67. Fort Ancient Jar (Rare), Ohio Area
$500.00

POTTERY PIPES

68. Human Effigy, 4"
$1,200.00

69. Human Effigy, 3-1/2"
$950.00

70. Noded Rim, 2-1/2"
$375.00

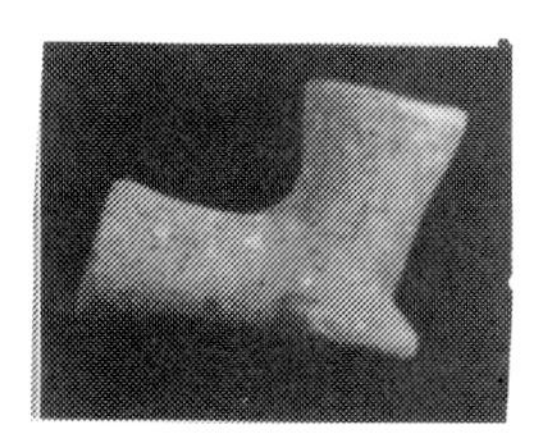

71. Elbow Pipe, 3-1/2"
$350.00

72. Archiac Tube Pipe - 6-1/2"
$750.00

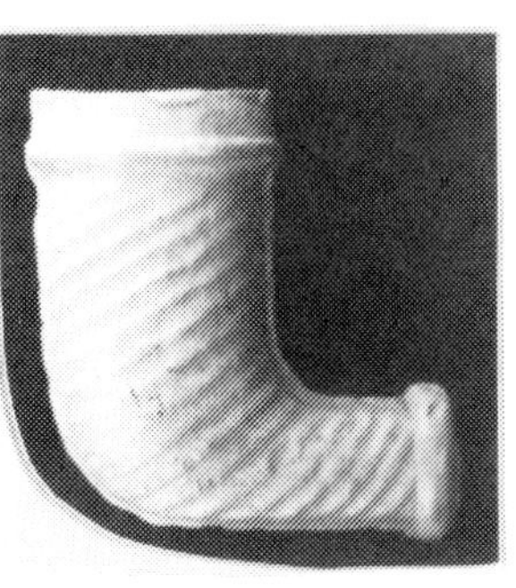

73. Trade Pipes, Various Types
$10.00 to $100.00

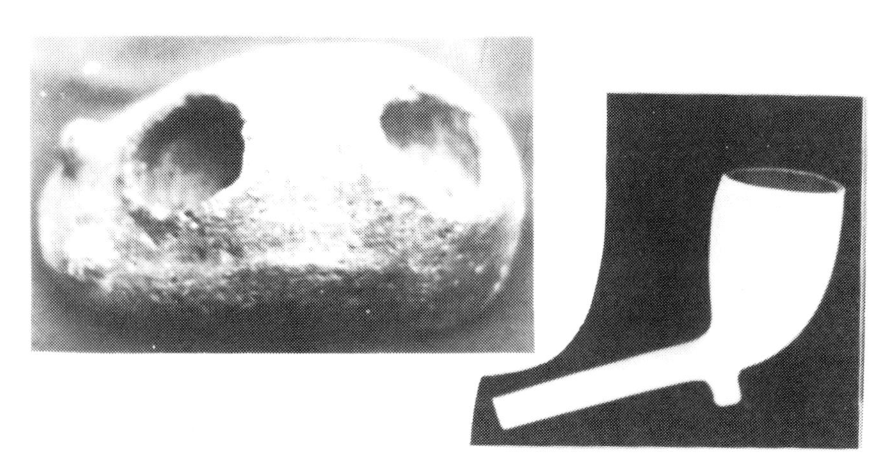

74. Frog Effigy, 4"
$950.00

75. English Trade Pipes
$35.00 to $300.00

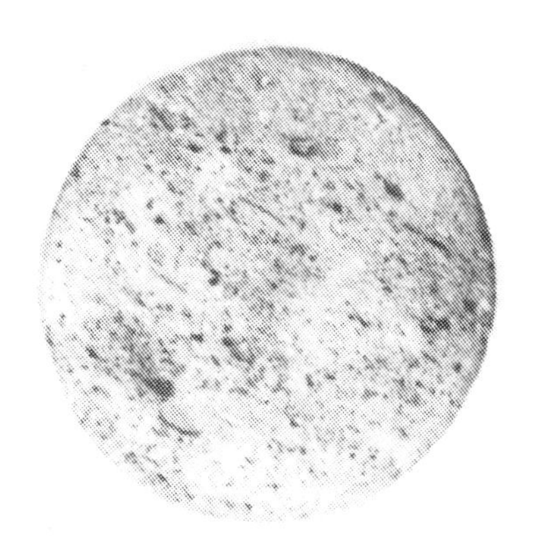

76. Pottery Sun Disc
3" Diameter
$200.00

SHELL

Shell was one of the Indians favorite materials. They used it to make tools, jewelry and utensils. It was also used in its natural state for spoons, cups, hoes, knives and hafted for clubs and picks.

Conch shells were used as drinking cups for the famous Black Drink ceremony. During this purification ceremony, the men (women were excluded) met on the village square and drank as much of the Black Drink as they could hold. Then they had a contest to see who could spew it the furthest when it came back up. This drink was brewed from the leaves of the holly tree. When it was taken, it caused nausea and a violent upheaval.

Conch and other large shells were also cut up and made into earpins, beads, pendants and gorgets. The gorgets and pendants were often engraved with cult or religious symbols.

Along the Atlantic coast, clam shells were cut up and made into small cylindrical beads, which were strung as necklaces or woven into belts. These belts were a form of money and were called Wampum by the Indians.

Small shells such as the Marginella were attached to clothing and capes. These often formed designs and sometimes even elaborate pictures.

In the Mound Building tribes, the most important use of shell was in the production of engraved disks called gorgets. These were highly polished and then engraved with cult designs. These represented rattlesnakes, spiders, birds, geometric designs, death masks, eagle dancers, bat wing dancers and other symbolic designs.

Fossil shells were also much prized by the Indians. They were used as charms and fetishes. They were placed on the altars in temples and in medicine bags.

Certain tribes in the interior seem to have placed a special significance on certain varieties of shell, especially those which came from the sea. They were buried with their dead. The value placed on beads made of shell even exceeded that of a human life at times. It has been reported that a murderer could buy atonement from the victims family by presenting them with a single string of well-crafted shell beads.

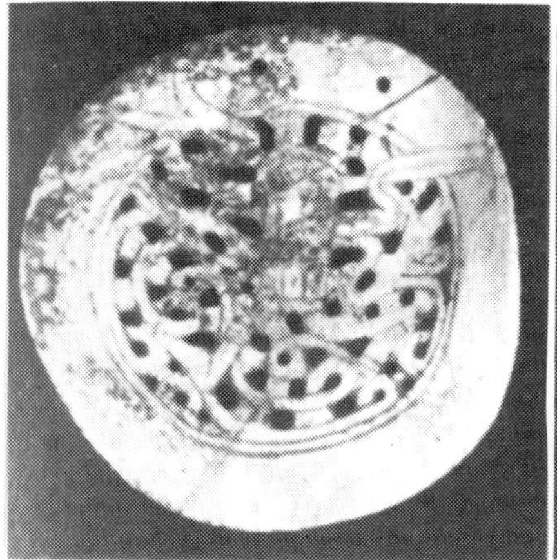

1. Dancing Bird Man
3" .. $5,000.00
4" .. $7,000.00
5" .. $9,500.00
6" .. $15,000.00

2. Rattle Snake, 6"
$7,500.00

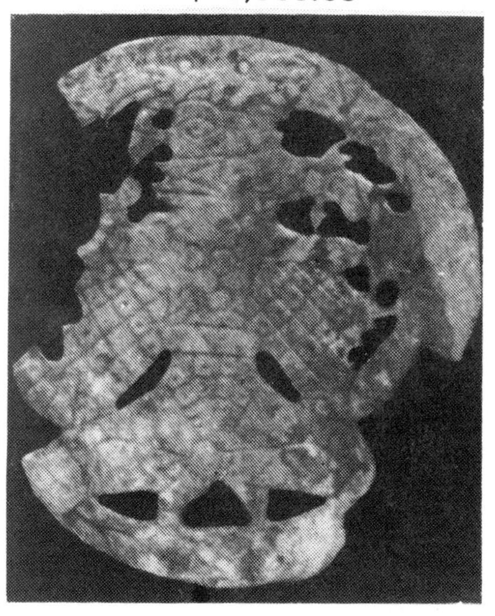

3. Eagle Dancer (Bat Wing Type) 4"
Intact .. $12,000.00
Damaged .. $5,000.00

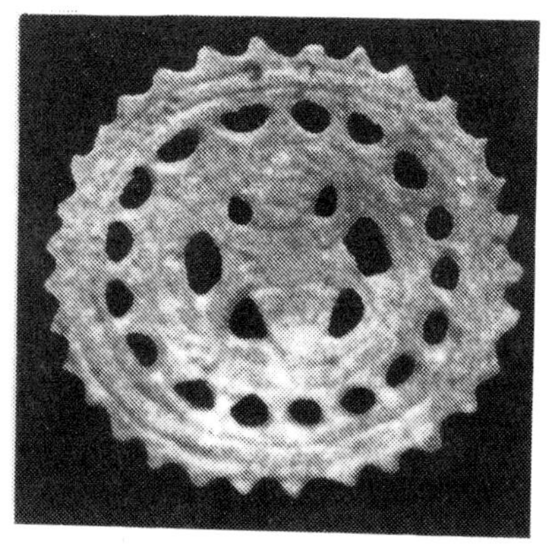

4. Spider, 2-1/2"
$5,000.00

5. Barred Oval Woodpecker, 2 Known to Exist, 3-1/2"
$8,000.00

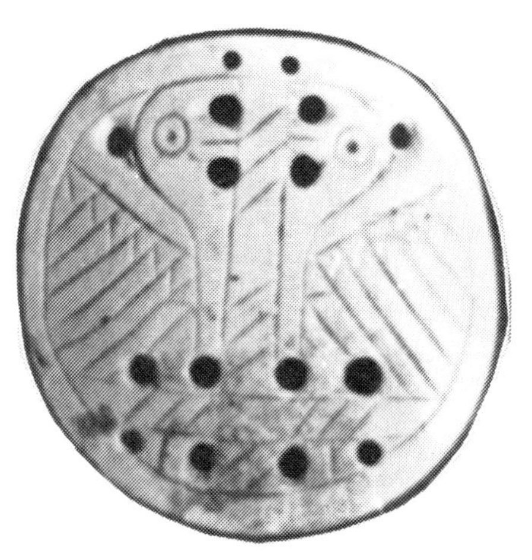

6. Double Turkey Cock 2-3/4"
$5,000.00

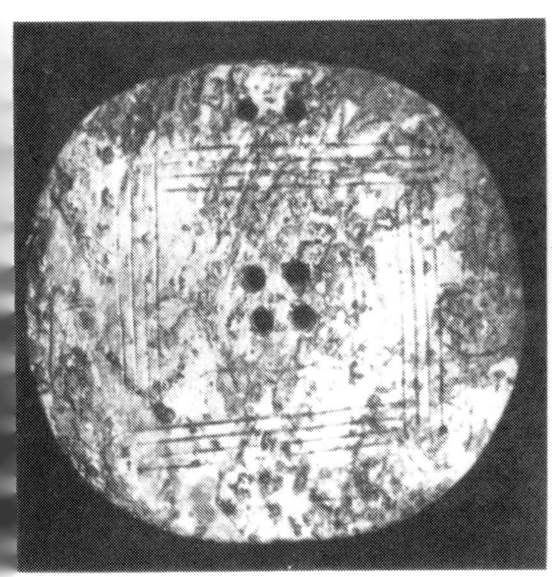

7. Woodpecker, 3"

$3,500.00

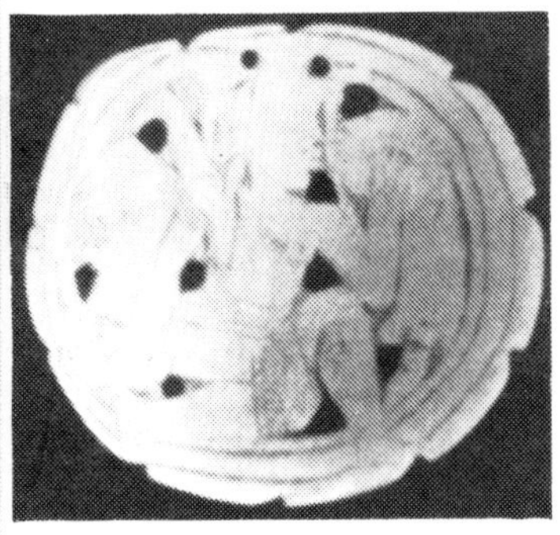

8. Warrior with Turkey and Raccoon, 4"
$10,000.00

9. Rattle Snake, 4"

$3,500.00

10. Death Mask, Rattlesnake on back, 6"

$7,500.00

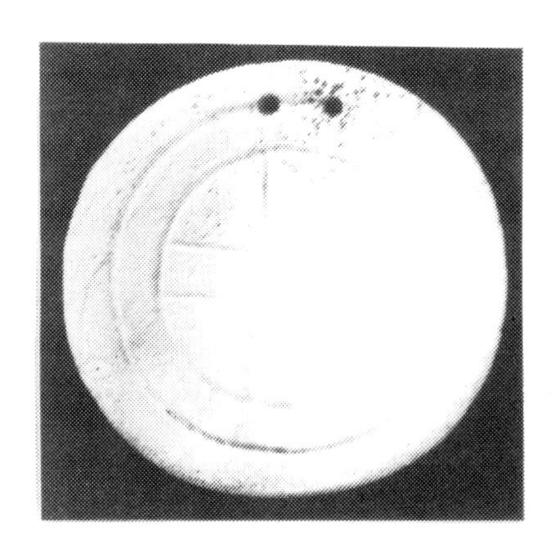

11. Cross and Sun Disc, 3"

$2,500.00

12. Sunburst, 2-1/2"

$2,500.00

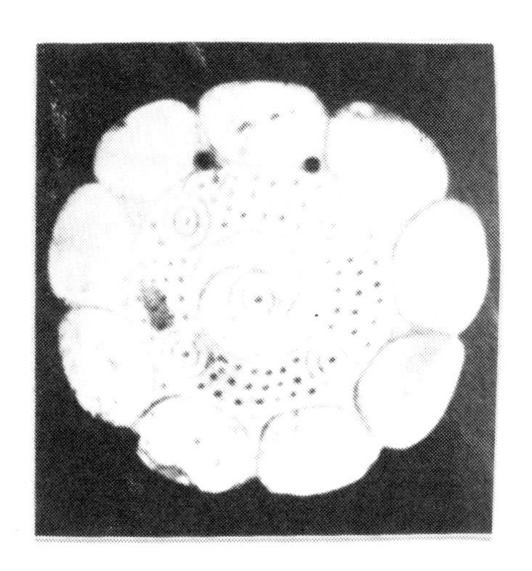

13. Triscal, 3-1/2"
$3,500.00

14. Triscal, 3"
$3,000.00

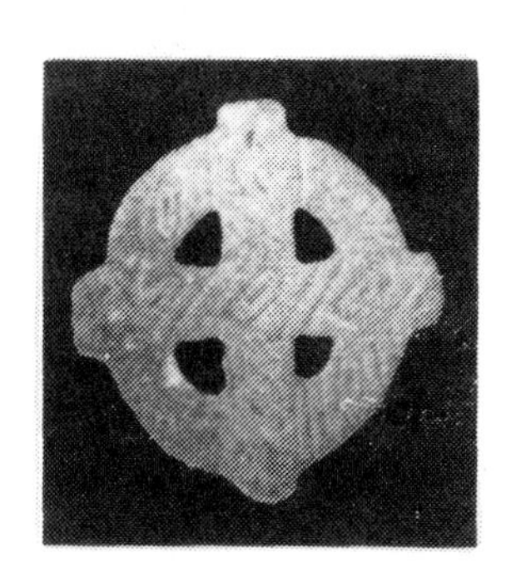

15. Barred Oval Cross, 2"

$2,000.00

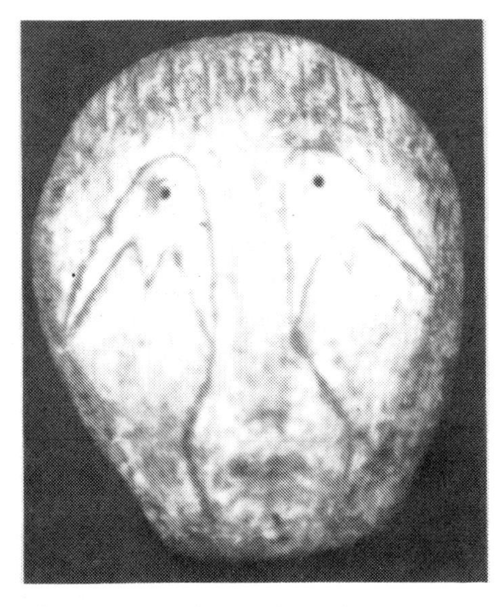

16. Death Mask with Weeping Eyes (actually the weeping eyes are cranes), 6"
$4,500.00

17. Double Birdman, Unusual Type, 3"
$4,500.00

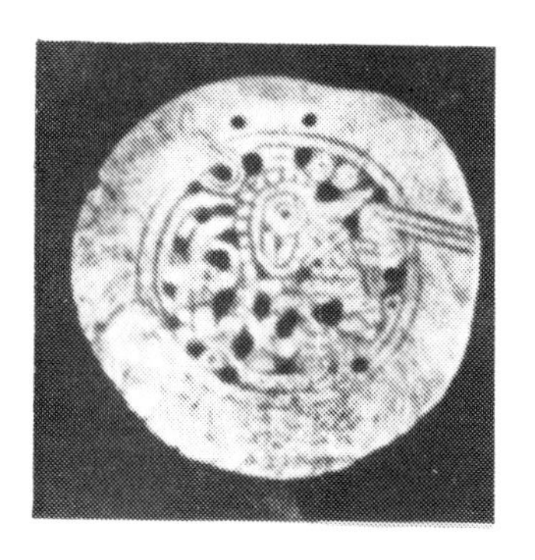

18. Dancing Birdman, 4"
$5,000.00

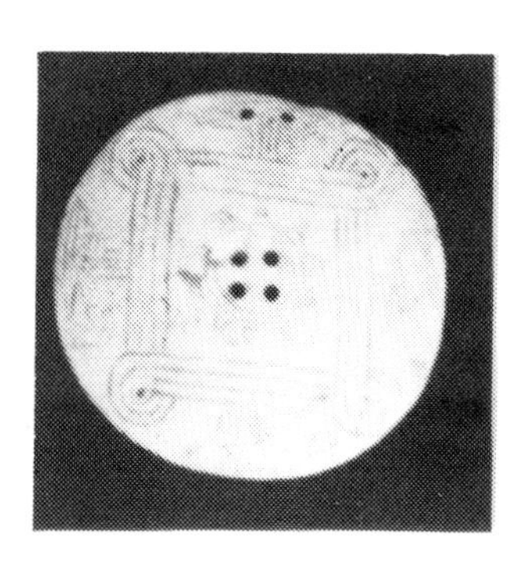

19. Woodpecker 4"
$7,500.00

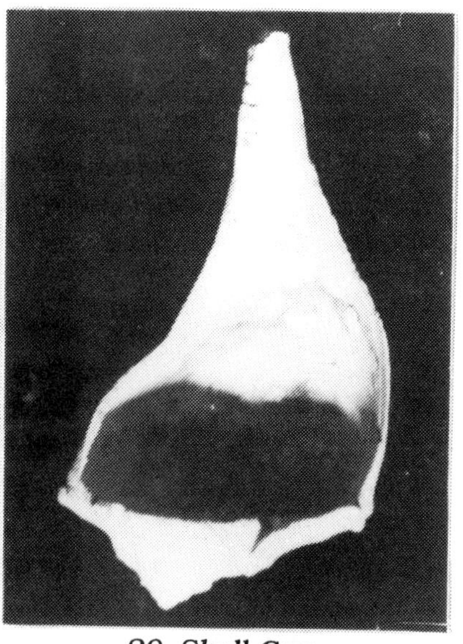

20. Shell Cup
14" .. $500.00
10" .. $375.00
8" .. $250.00
4" .. $200.00

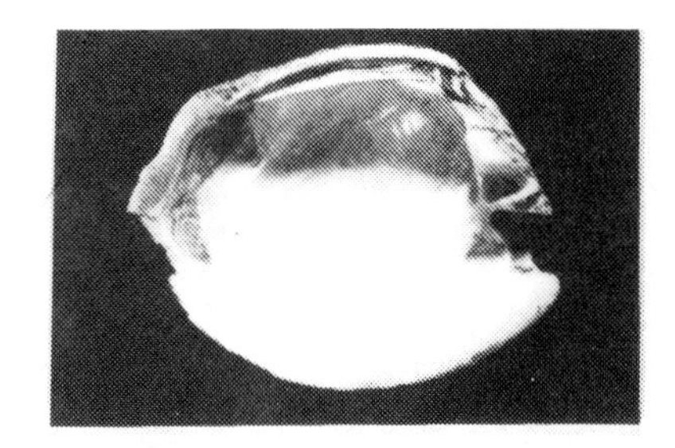

21. Shell Spoon
$50.00

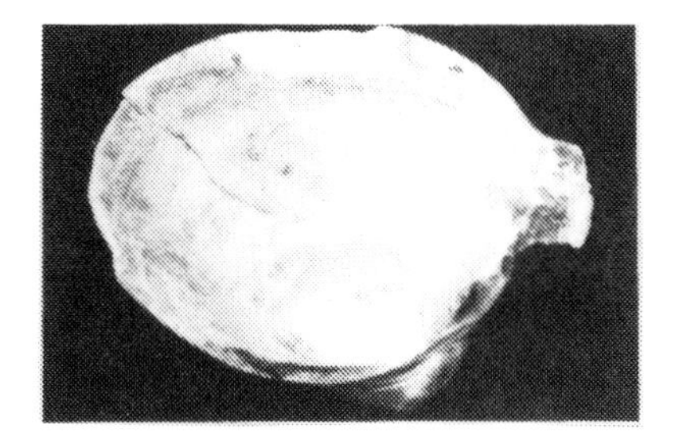

22. Shell Spoon-Carved Handle
$100.00

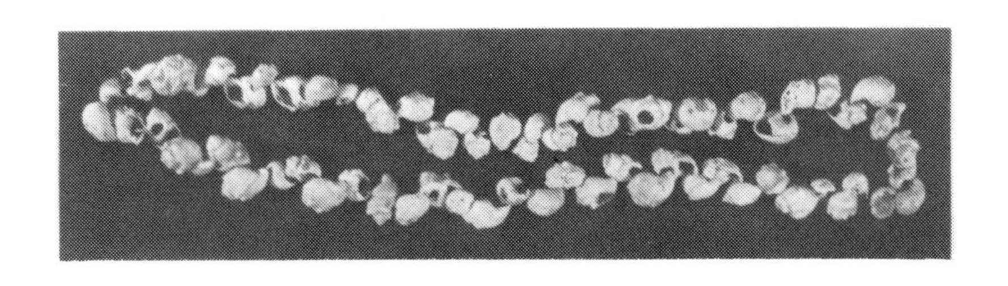

23. Shell Beads, 22"
$125.00

28

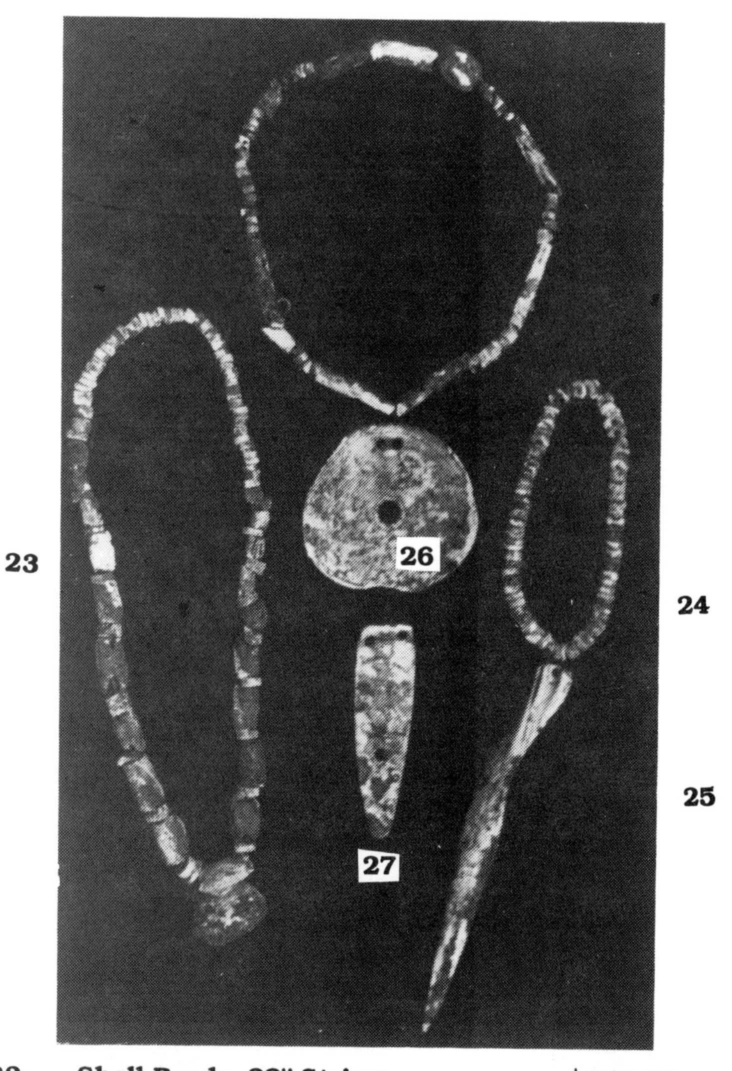

23

24

25

23.	Shell Beads, 22" String	$350.00
24.	Shell Beads, 10" String	$100.00
25.	Shell Columella Pendant, 6"	$375.00
26.	Sun Circles Gorget, 3"	$1,500.00
27.	3-Hole Pendant, 4"	$375.00
28.	Shell Beads, 18" String	$200.00

STONE

Early man first learned to fracture stone in order to manufacture scraping and cutting tools. Later, he learned to flake and grind stone into more useful shapes. This was not only to increase the convenience and effectiveness of his tools, but also to add beauty to those tools.

Man is not the only creature who uses tools, but he is the only one who goes to the trouble of making them into works of art.

The Indians of north America were still in the Stone Age when contact was made with the Old World. Metal had been used since the Archaic period. This was in the form of native copper which was found in an almost pure state. Its use was very limited in area and scope.

The Indians used all workable types of stone that were to be found in their territories and also imported special types of stone from outside areas. Sometimes these special types of stone were carried vast distances from one area to another. Pipe stone from the northern United States was traded to the Gulf coast. Obsidion from the Rocky Mountains was traded as far east as Ohio.

These materials were processed by using one or more of the following procedures: fracturing; pecking; incising or cutting; abrading and drilling; and polishing. At one time these crafts were thought to be lost. Today, however, there are numerous individuals who are highly skilled in these procedures. Some are skilled artists whose work is marked and sold as reproductions. Unfortunately, many more are unscrupulous scoundrels whose only goal is to separate the uneducated from their money. In case the reader is one who thinks this problem is a recent occurence, the following quotation may be enlightening.

"The knowledge acquired in recent years through experiments in stone shaping processes has led unfortunately to the manufacture of fradulent imitations of aboriginal implements and sculptures for commercial purposes, and so great is the skill acquired in some cases, that it is exceedingly difficult to detect the suspicious work; there is thus much risk in purchasing objects whose pedigree is not fully ascertained. See "Pseudo Indian".

This was printed in Bulletin 30 of the Smithsonian Institution entitled Handbook of the American Indian, printed in 1912.

BOAT STONE

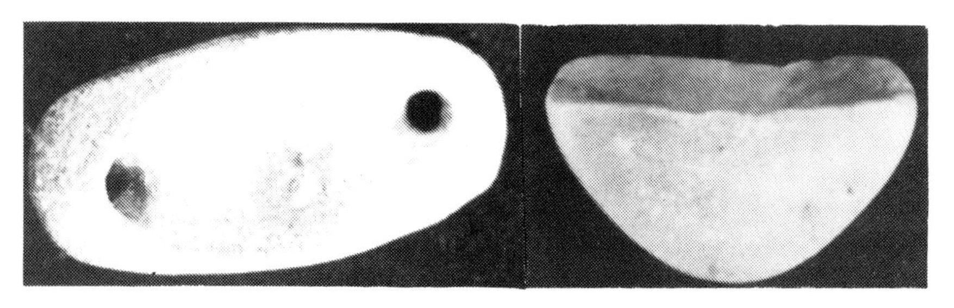

1. *Limestone*
$75.00

2. *Polished Quartz*
$175.00

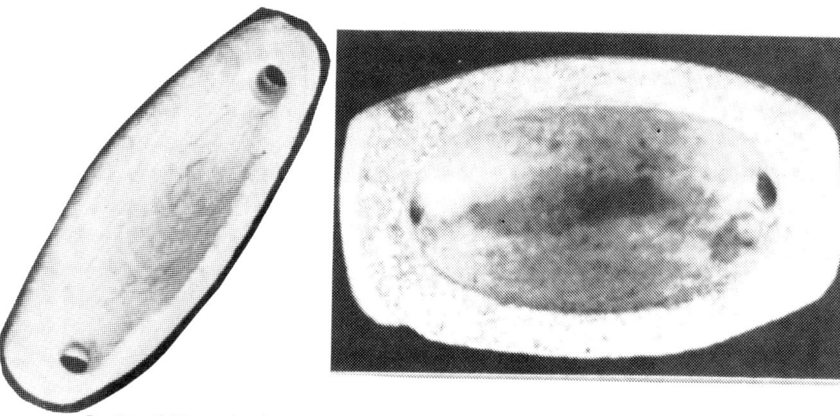

3. *Red Banded Slate*
$100.00

4. *Deep Scooped, Nice Finish*
$350.00

5. *Polished Quartz, Deep Scooped*
$500.00

6. *Top View of Quartz*

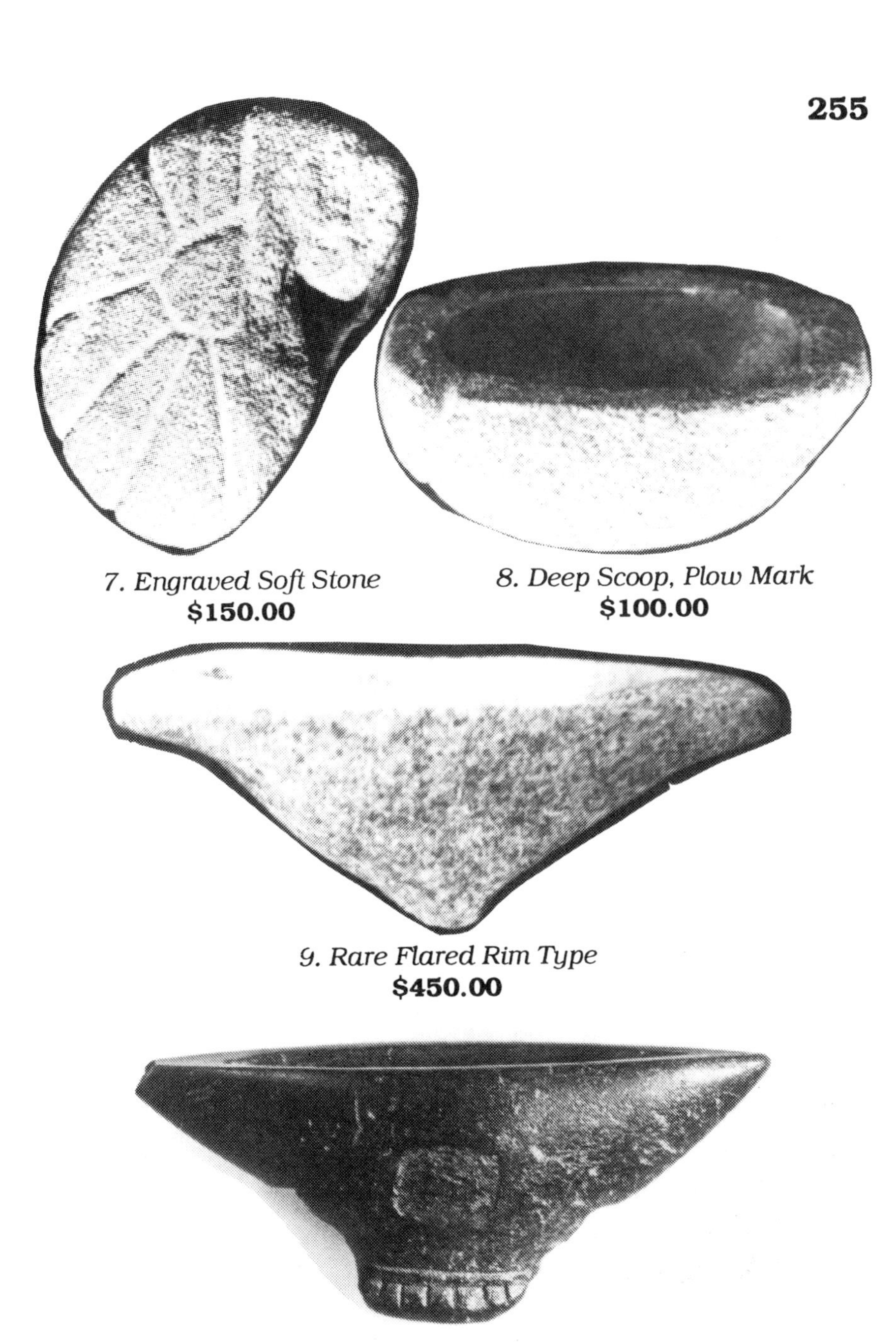

7. *Engraved Soft Stone*
$150.00

8. *Deep Scoop, Plow Mark*
$100.00

9. *Rare Flared Rim Type*
$450.00

10. *Keeled Boat Stone, possibly had an inlayed side and engraved, very rare, slight damage*

$750.00

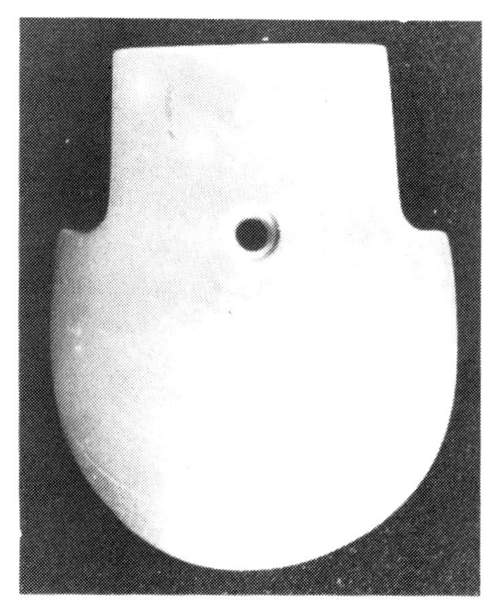

11. Spatulate Spud

$1,500.00

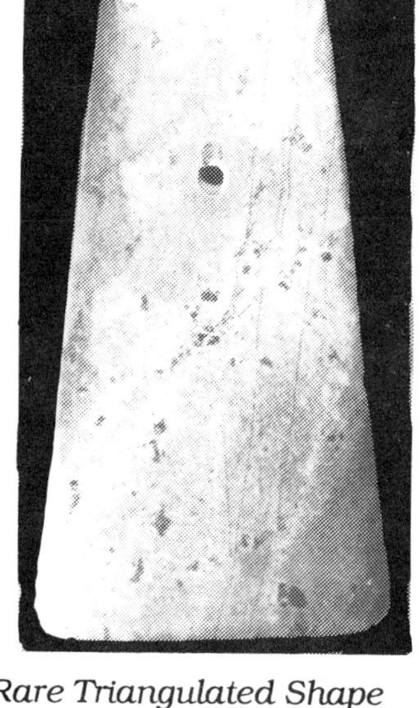

12. Rare Triangulated Shape Black Veined Green Slate

$2,000.00

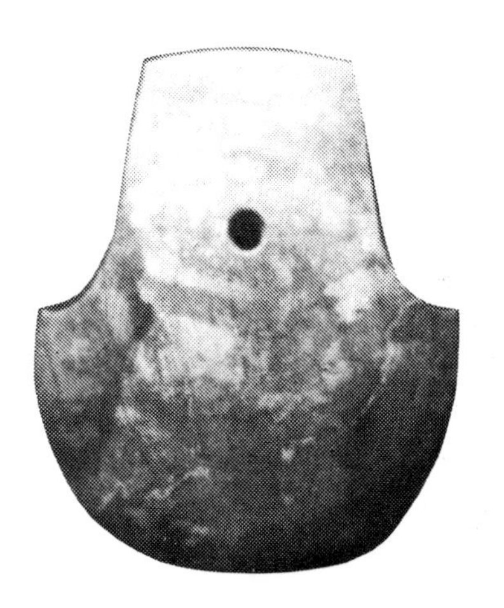

13. Approximately 6" Long

$2,500.00

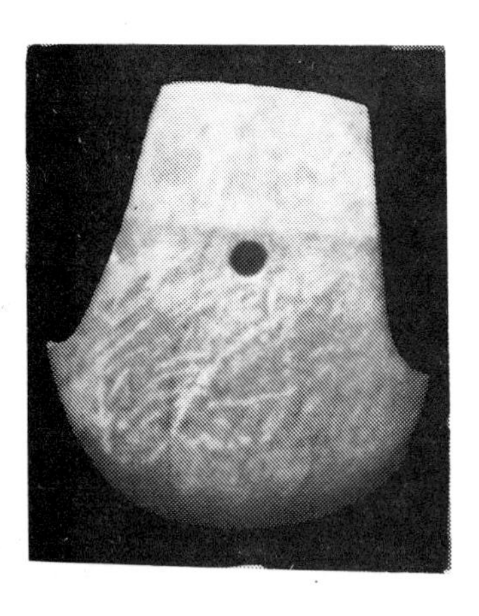

14. Approximately 4.5"

$1,200.00

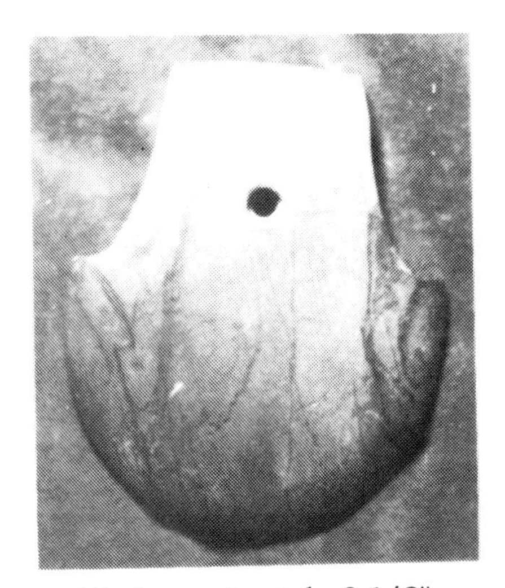

15. Approximately 6-1/2"
$1,700.00

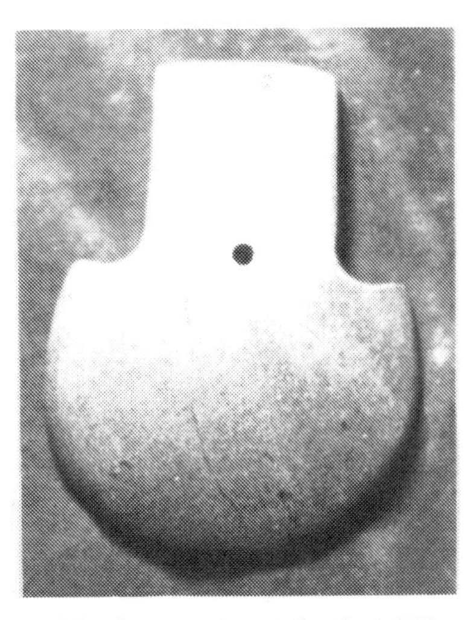

16. Approximately 6-1/2"
$1,800.00

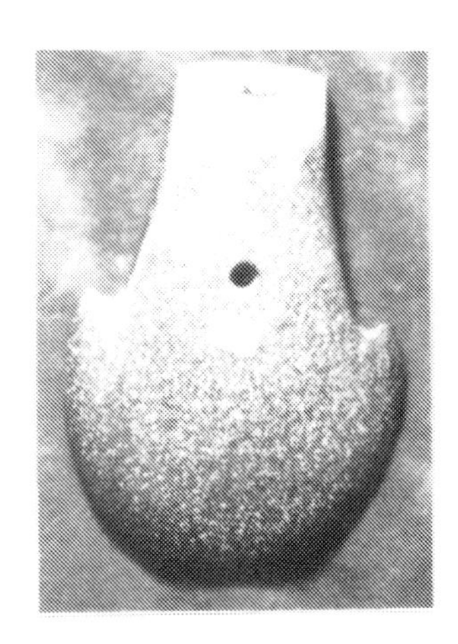

17. 5"
$1,200.00

18. 7"
$1,900.00

19. Broom Handle Spud
Broken, 8" .. $600.00
Unbroken .. $3,500.00

CELT

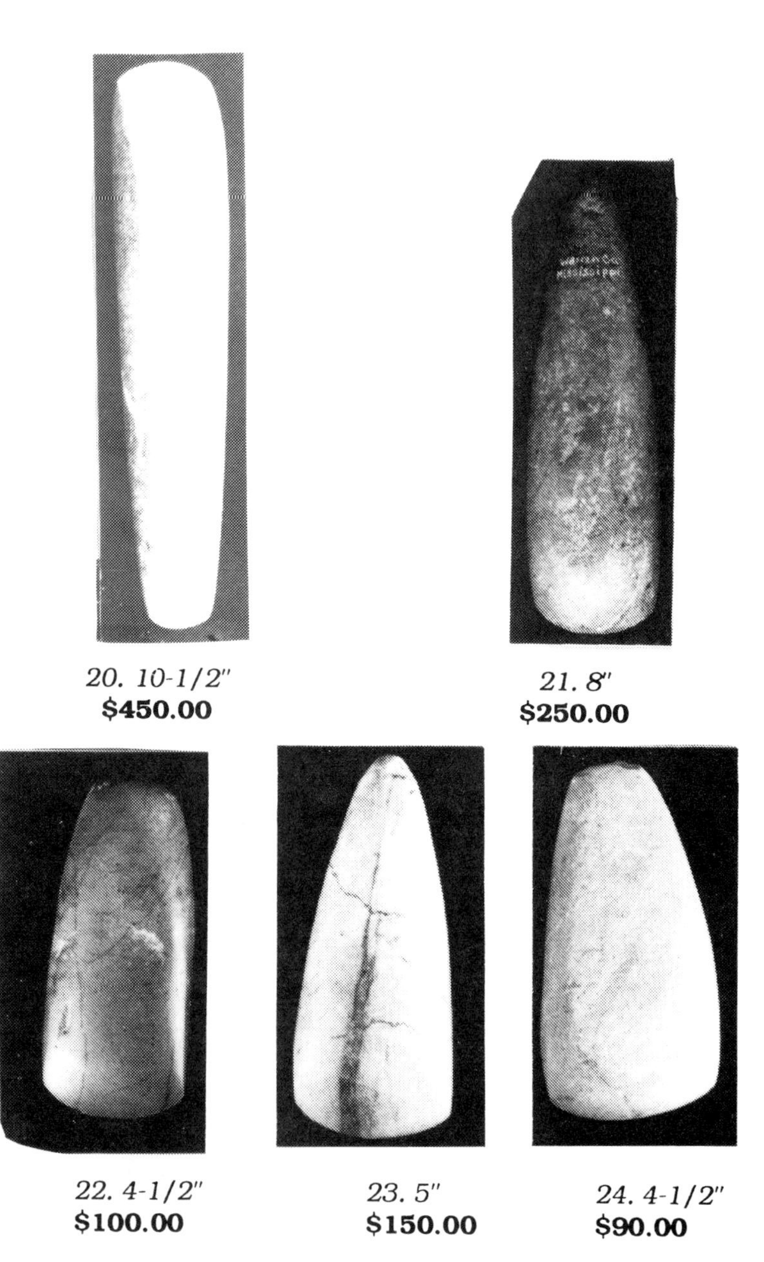

20. 10-1/2"
$450.00

21. 8"
$250.00

22. 4-1/2"
$100.00

23. 5"
$150.00

24. 4-1/2"
$90.00

MINIATURE CELTS

Size reduced 25%

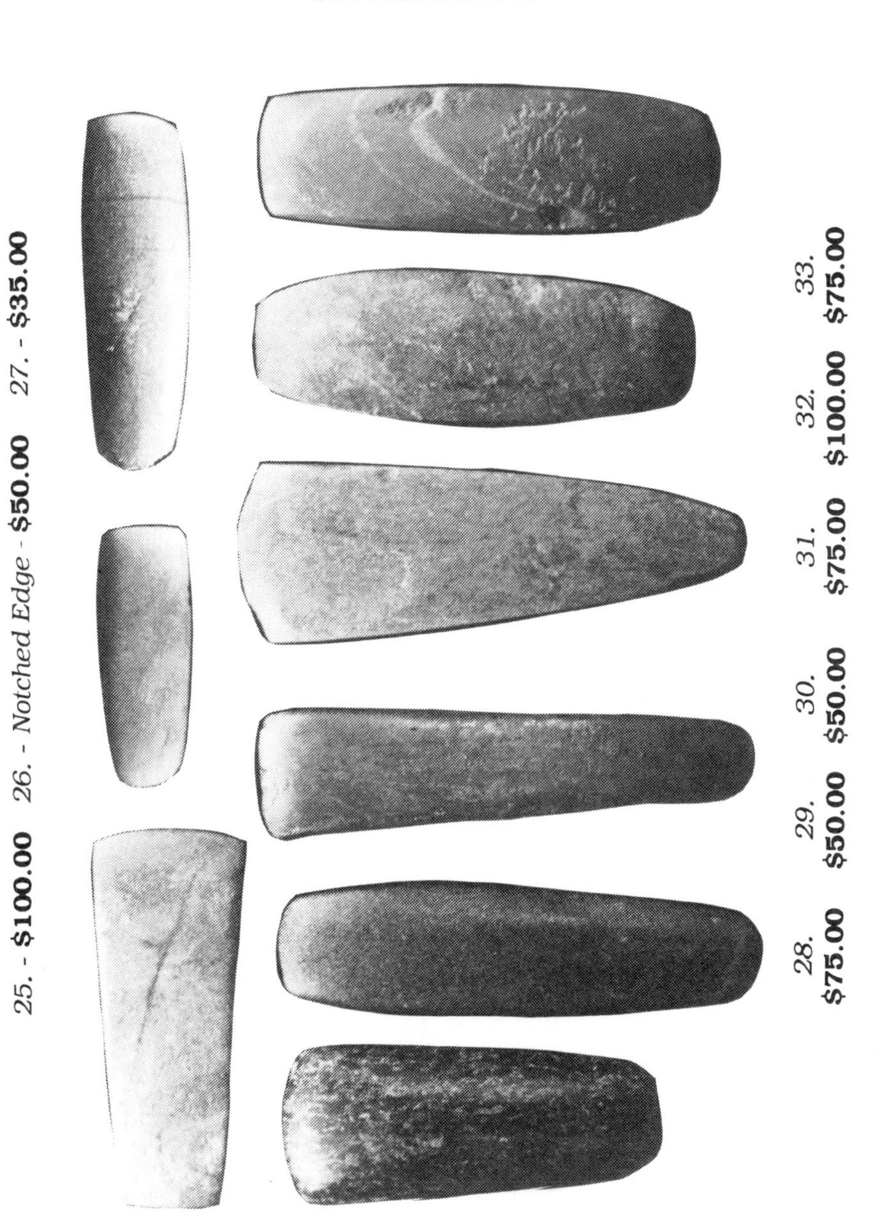

25. - **$100.00** *26. - Notched Edge* - **$50.00** *27.* - **$35.00**

28. **$75.00** *29.* **$50.00** *30.* **$50.00** *31.* **$75.00** *32.* **$100.00** *33.* **$75.00**

STONE BEADS

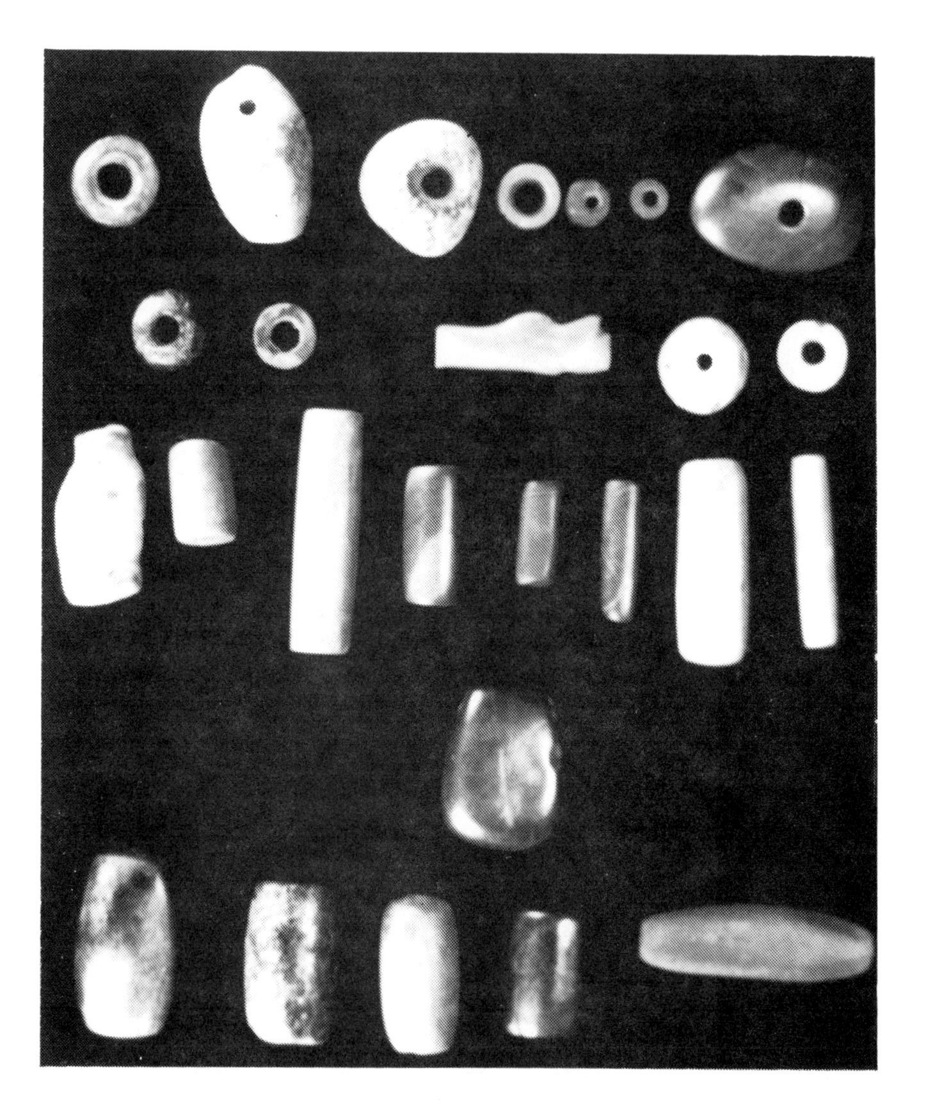

34. TOP ROW - Left to right
$25,00, $45.00, $15.00, $5,00, $5.00, $5.00, $45.00
2nd Row: **$20.00, $15.00, $350.00, $15.00, $15.00**
3rd Row: **$75., $45., $200., $175., $45., $75.0 $75., $125.**
4th Row: **Restored Pendant Type .. $45.00**
5th Row: ***Drilling Started $15., $100., $50., $150., $100.***

PICK

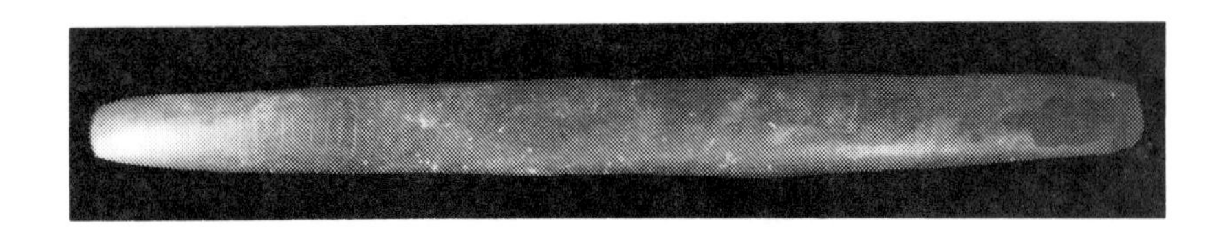

35. Engraved Pick, 10-1/2 inches
$1,500.00

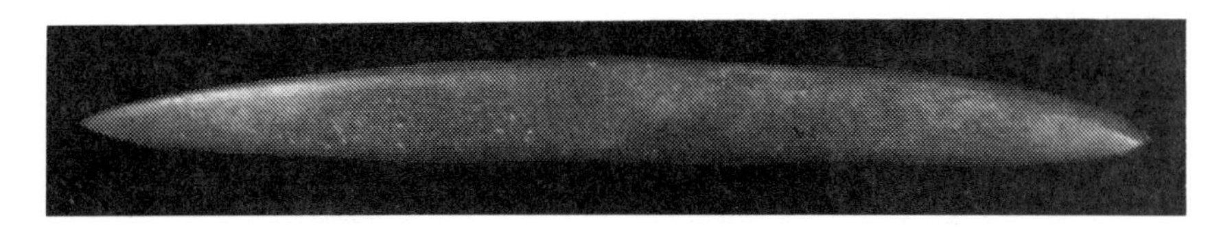

36. Pick, 10 inches
$750.00

GROOVED AX

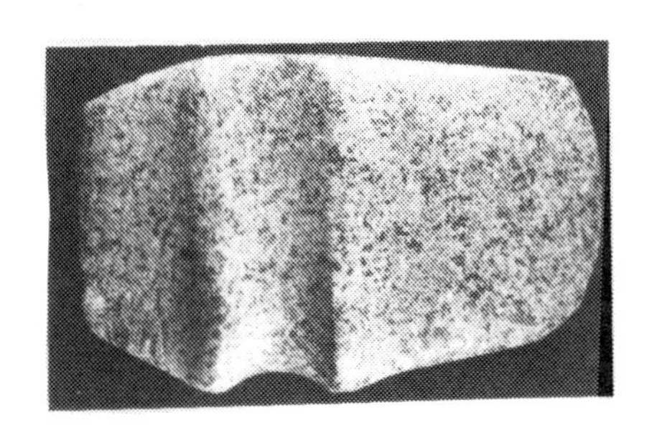

37. 3/4 Ground Ax - Raised Barbs Fluted Top and Bottom, 5-1/5 inches
$1,000.00

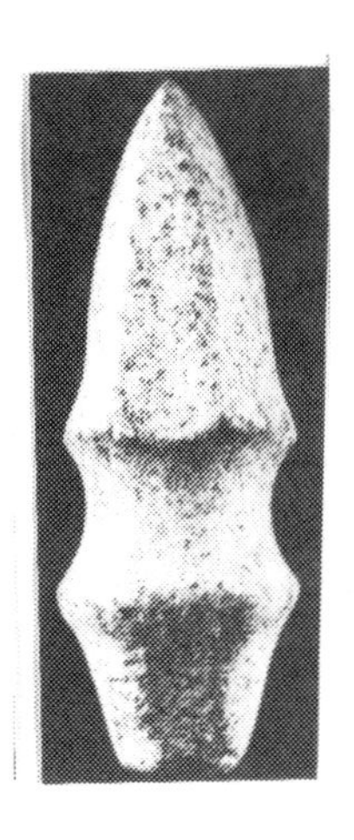

38. Top View of Left

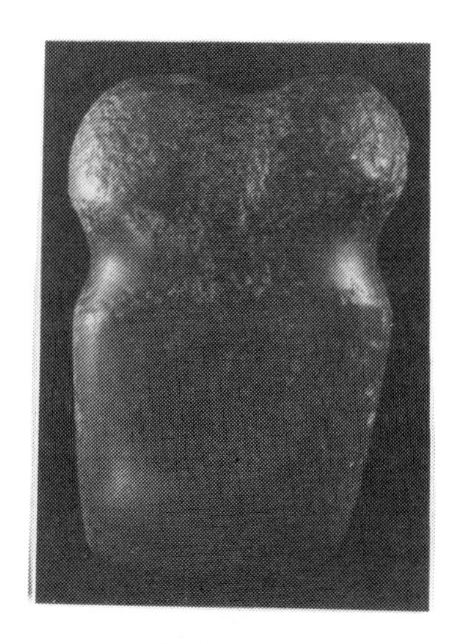

39. Rare Rabbit Earred Ax
Red Hematite, 6-1/2"
$2,000.00

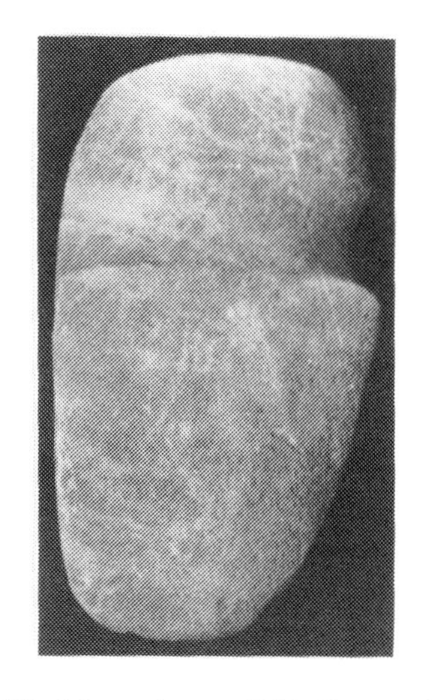

40. Limestone, 3/4 Ground
6"
$175.00

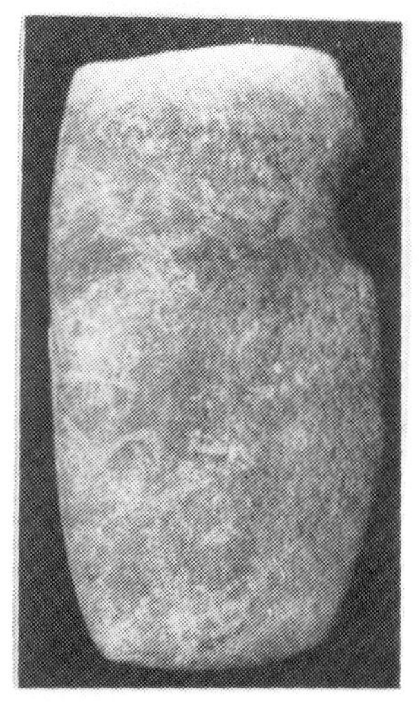

41. Limestone, Rough Finish

$125.00

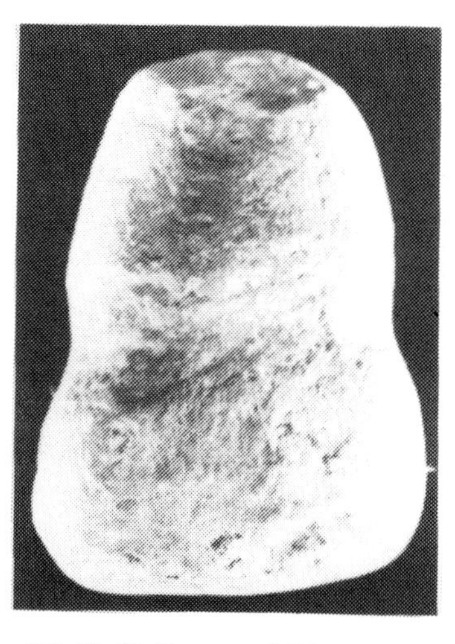

42. Full Grooved Hammer,
Limestone
$75.00

43. Southern Barbed Ax, Greenstone, 6-1/2"
$1,200.00

44. Deptford Ax, 8"
$1,000.00
(Material is Oorilite)

45. Deptford Ax, 8"
$1,500.00

CONGLOMERATE AX

46.

A. 5" Chipped Conglomerate
$150.00

B. 6-1/2" Chipped Celt Conglomerate
$200.00

C. 5" Chipped Ax Conglomerate
$50.00

GORGET

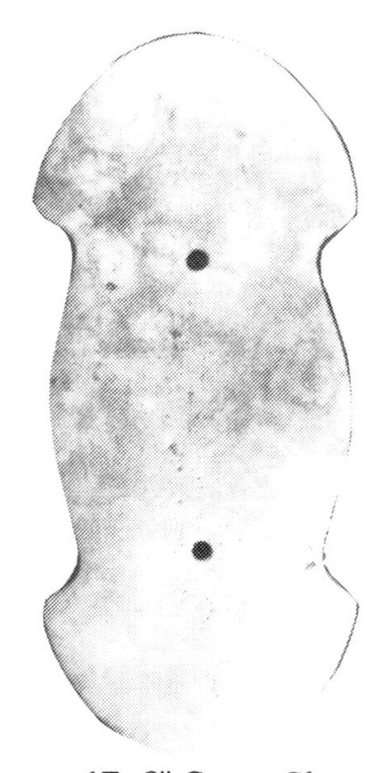

47. 6" Green Slate
$1,500.00

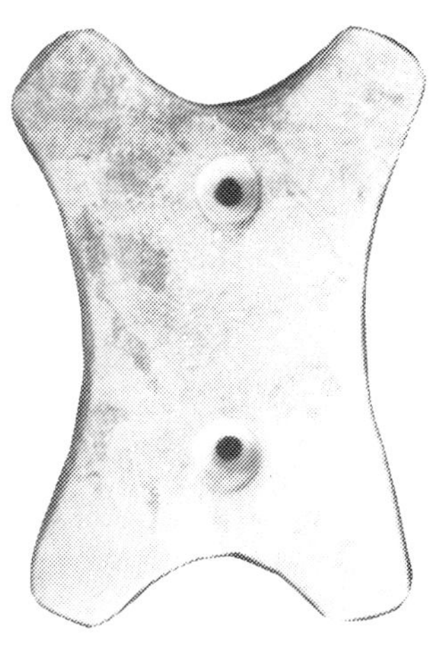

48. Reel Type Gorget, 4"
$500.00

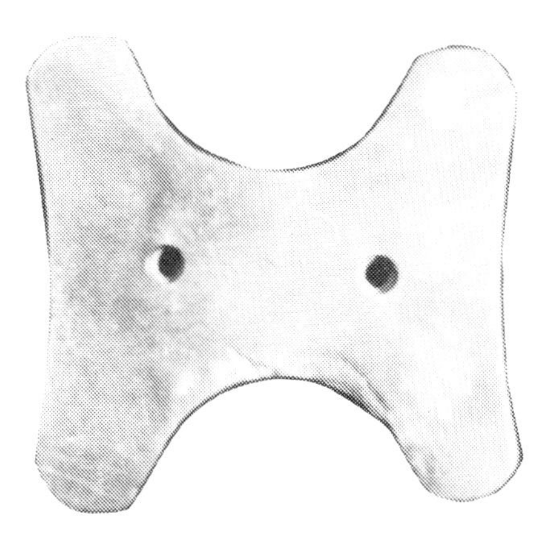

49. *Reel Type, 4"*
$600.00

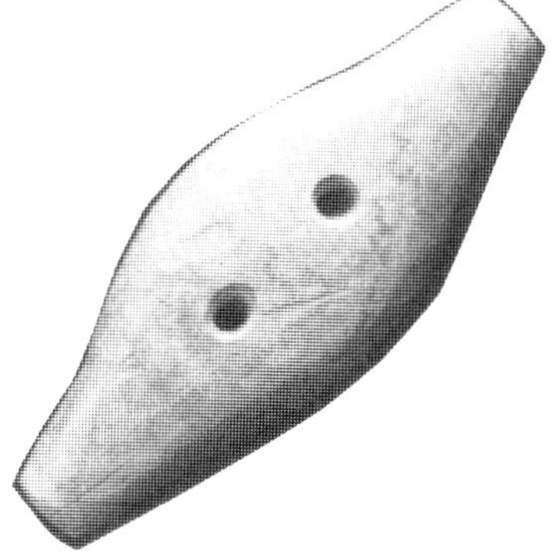

50. *Expanded Center Gorget, 4"*
$300.00

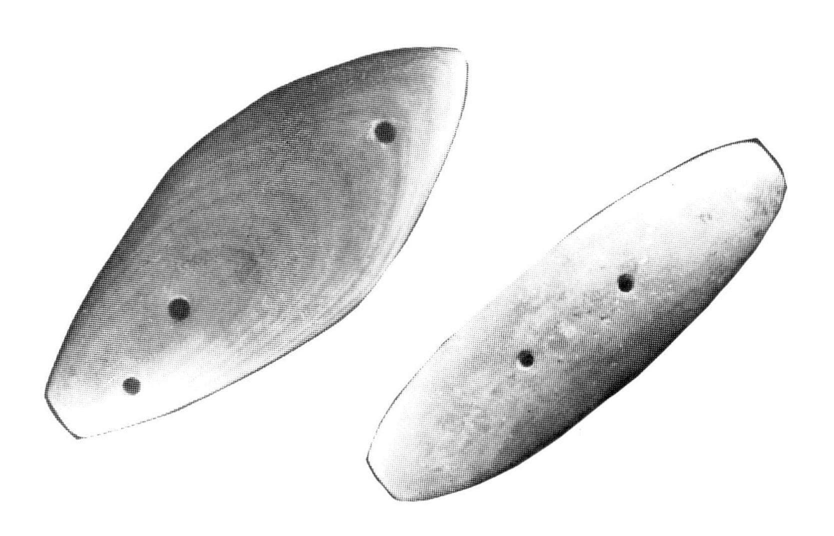

51. *Glacial Kame, 6"*
Red Banded Slate
$1,500.00

52. *Adena Gorget,*
6-1/2"
$500.00

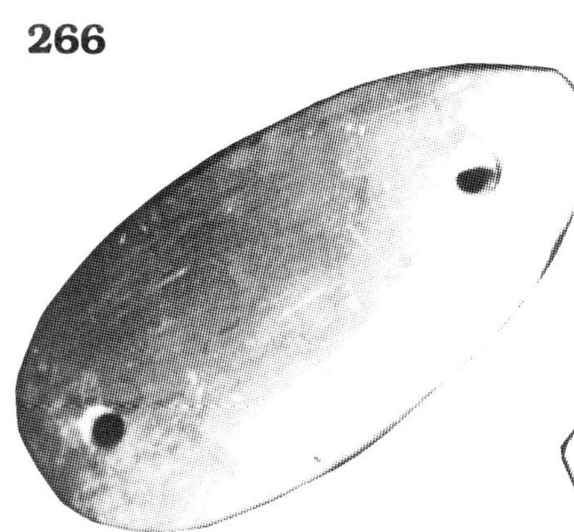

53. Glacial Kome, Red Slate

$400.00

54. Expanded Center, Red Banded Slate

$1,000.00

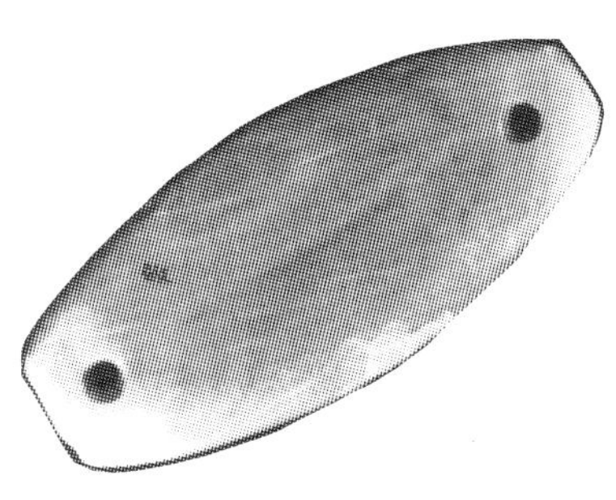

55. Red Banded Slate, 5"

$400.00

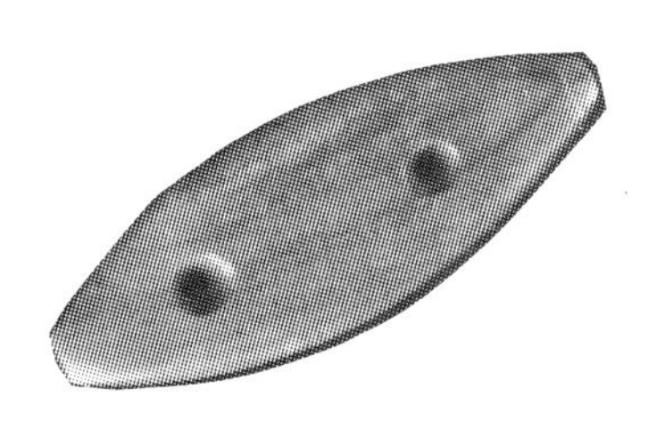

56. Adena Gorget, 4", Red Slate

$200.00

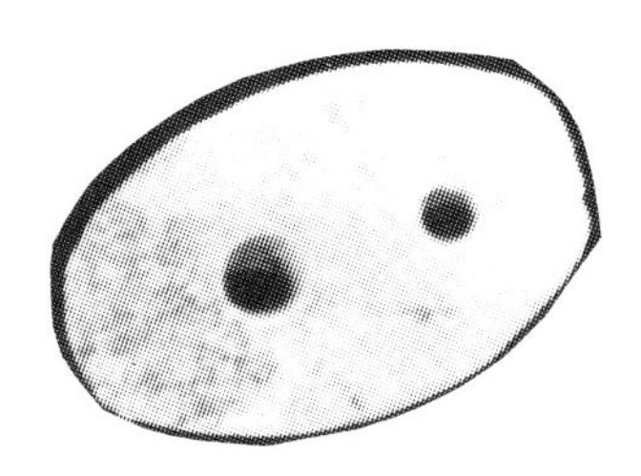

57. Adena Gorget, 2-1/2", Limestone

$100.00

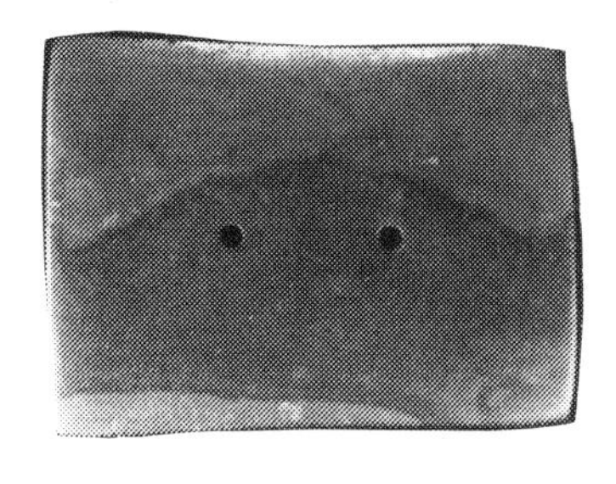

58. Adena Tablet Gorget Talley Marks on Edge, 4"

$1,000.00

PENDENT

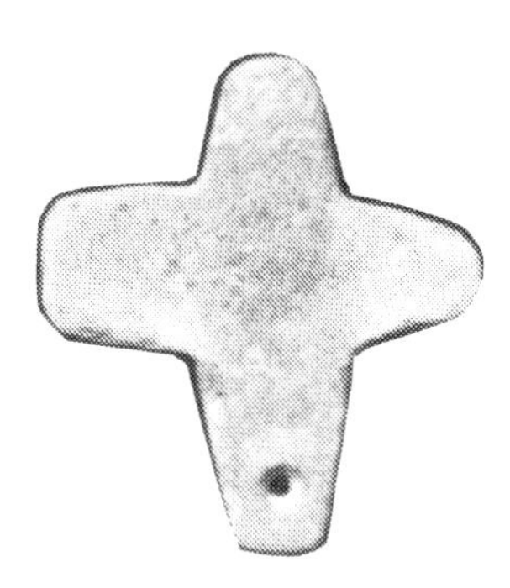

59. Monolithic Ax Pendent
$1,200.00

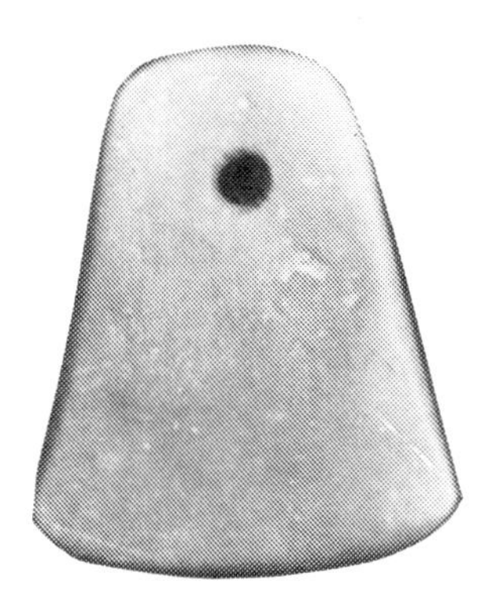

60. Red Slate
$125.00

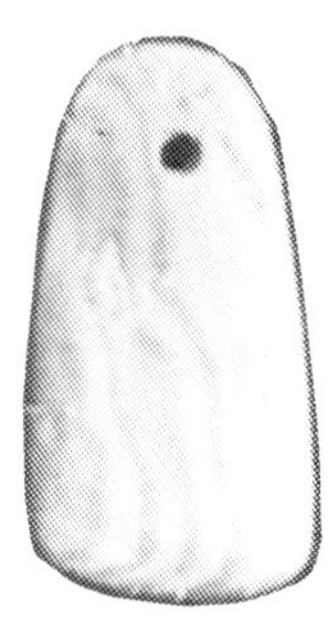

61. Red Banded Slate, Talley Marks
$200.00

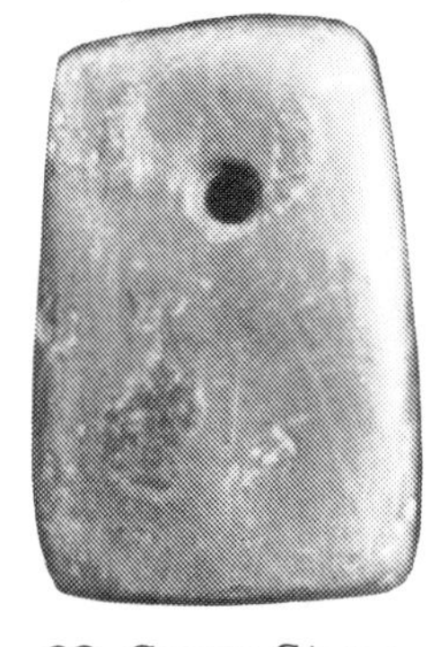

62. Green Stone

$75.00

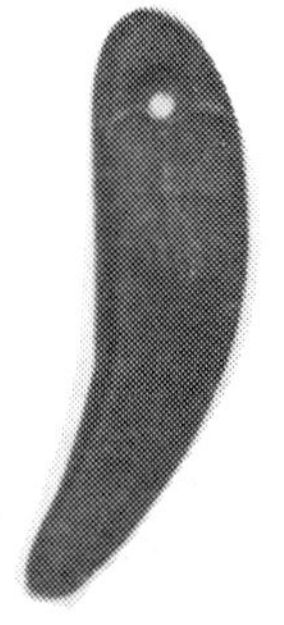

63. Bear Claw Effigy, Cannel Coal
$250.00

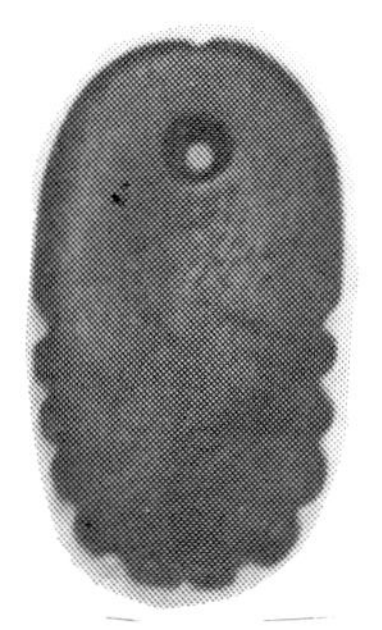

64. Notched Edge, Cannel Coal
$175.00

BANNER STONE

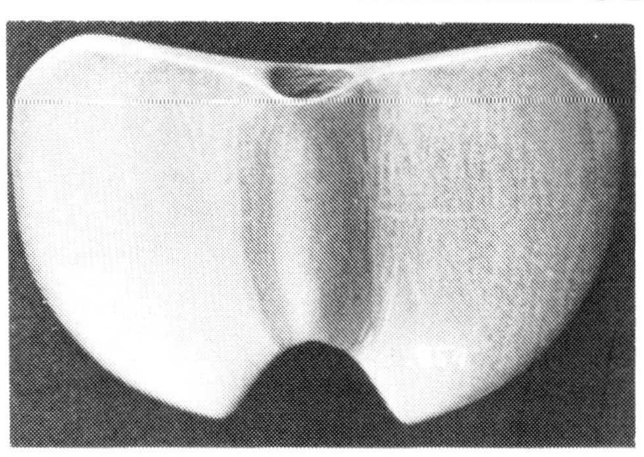

65. *Butterfly, Green Slate*
$2,500.00

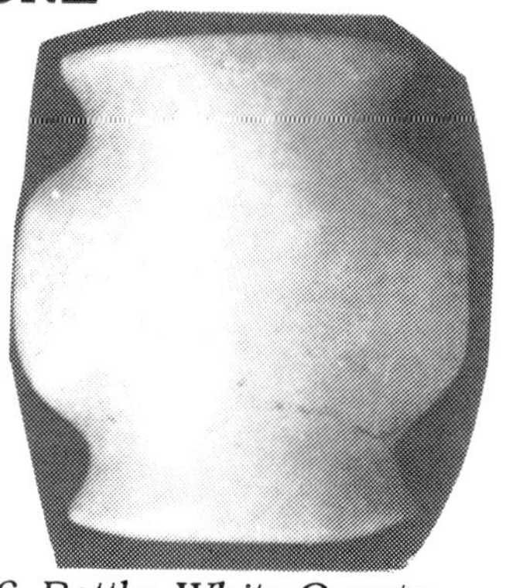

66. *Bottle, White Quartz*
$3,000.00

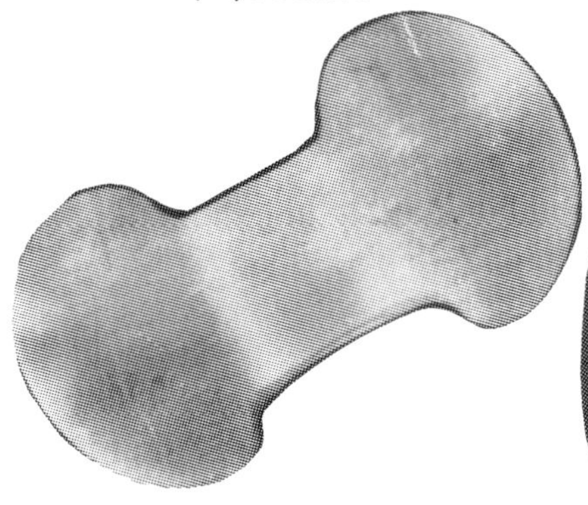

67. *Butterfly, Fergins Quartz*
$3,500.00

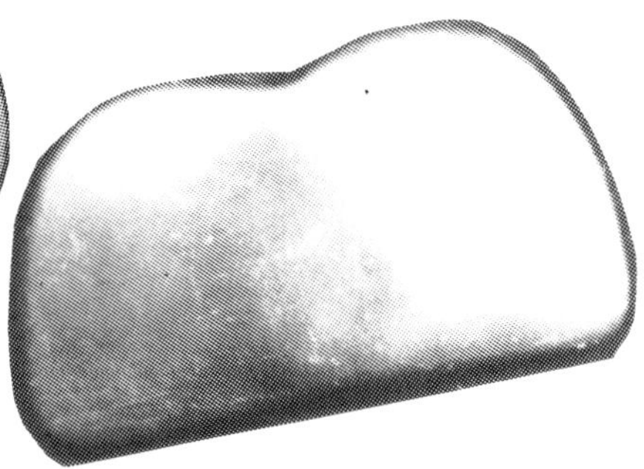

68. *Butterfly Type, Greenstone*
$800.00

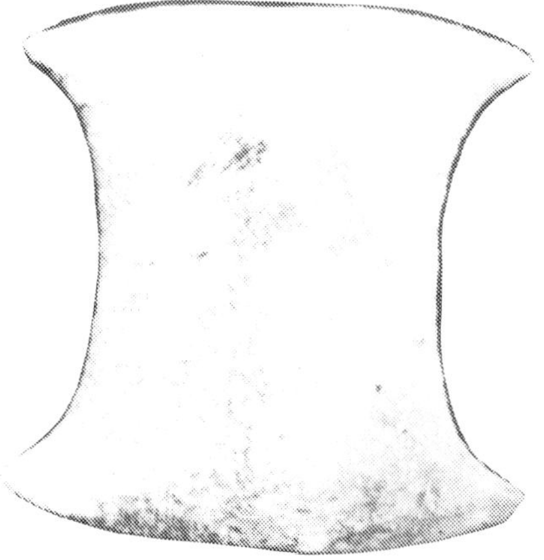

69. *Hourglass, Green Quartz*
$1,000.00

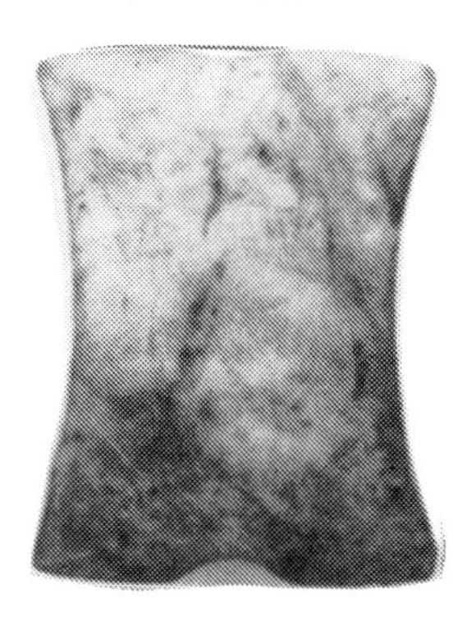

70. *Hourglass, Green Quartz*
$1,200.00

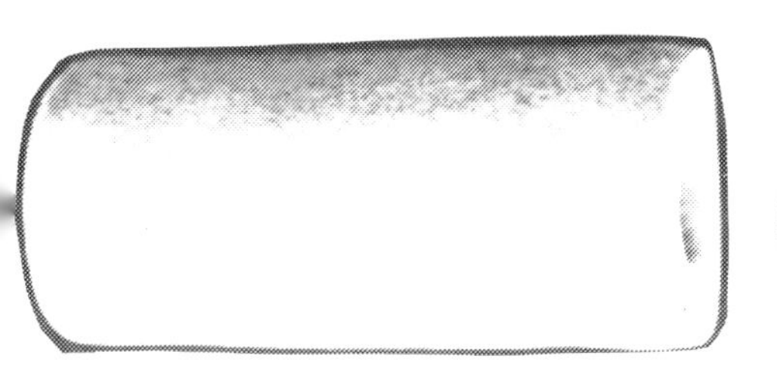

71. D. Hump. Banner, Red Slate
$700.00

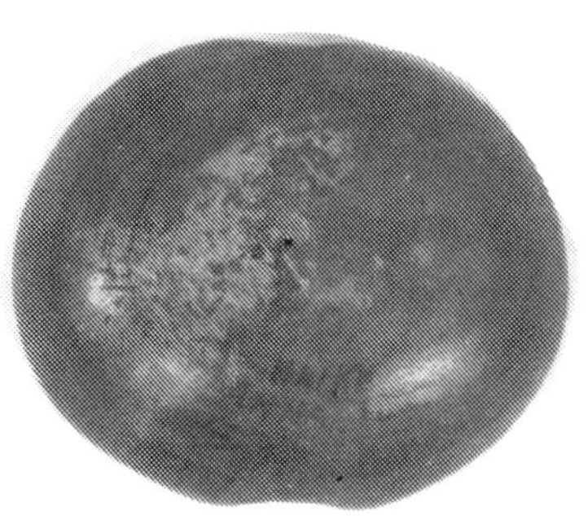

72. Ball Banner, Green Slate
$600.00

73. Football - Shuttle, Red Slate

$750.00

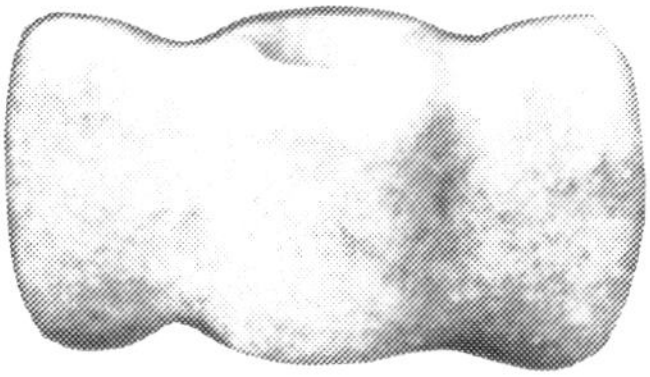

74. Southern Sugar Loaf Type, Rose Quartz
$900.00

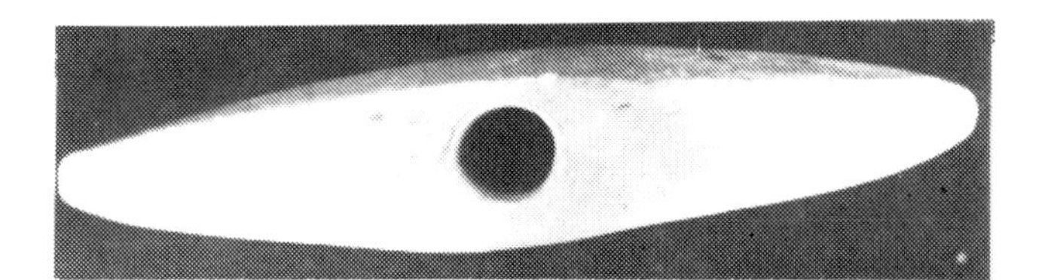

75. Shuttle Type, Greenstone
$450.00

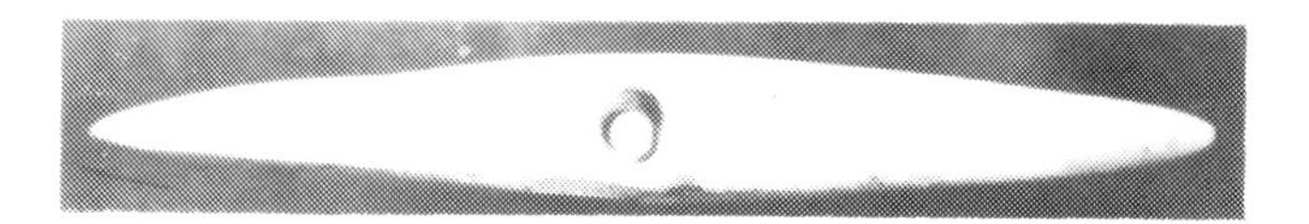

76. Shuttle Type, Partially Drilled (Rare) - 6"
$350.00

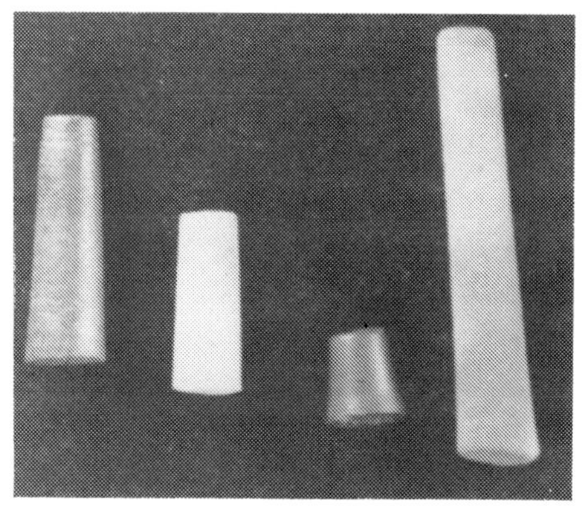

77. Cores from Banner Stone Drilling
$15.00 to $50.00

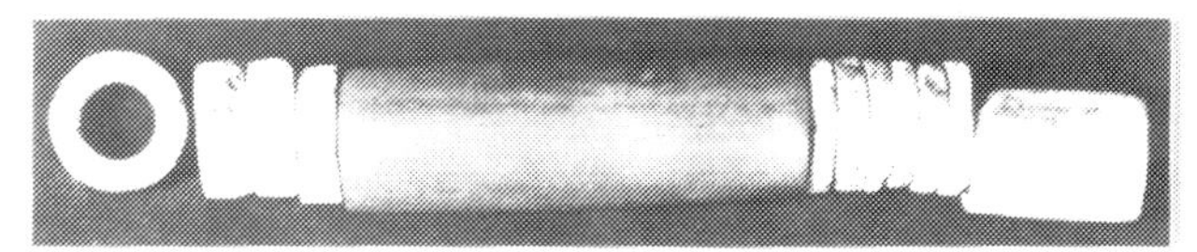

78. Composite Banner, Shell, Greenstone, Red Slate - 6-1/2"
$1,500.00

C B A
F E D

79.
Broken Stone Pieces. (A-F, Top right to left, bottom right to left)
A. Banner Half, $45.00 - B. Banner Half, $60.00 C. Gorget Half, $45.00 - D. Banner Half, $60.00 - E. Pipe Half, $100.00 F. Sun Disc Fragment, $75.00

HUMAN EFFIGYS, STATUS

80. Chlorite (Translucent Greens) 10" Tall, Museum Quality

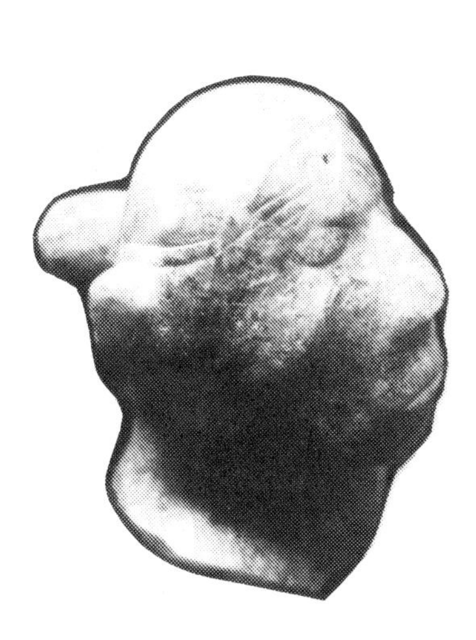

81. Reverse Side of Above

82. Sandstone Type Material, Approx. 18" Tall, Plow Damage
$12,000.00

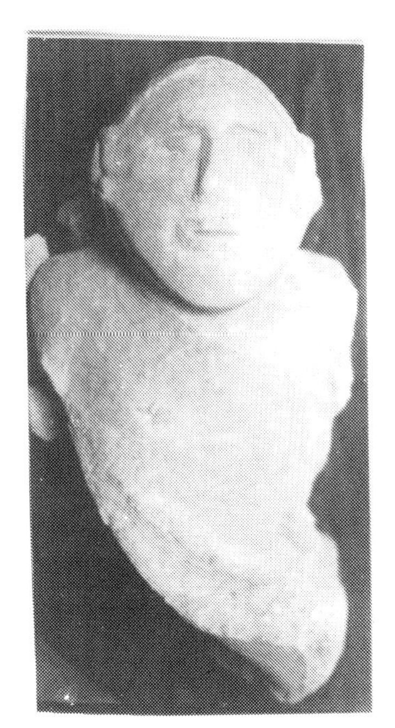

83. Sandstone Type Material, Unfinished, Approx. 20"
$12,000.00

84. Material Appears to be Pink Limestone, 3" Tall
$1,500.00

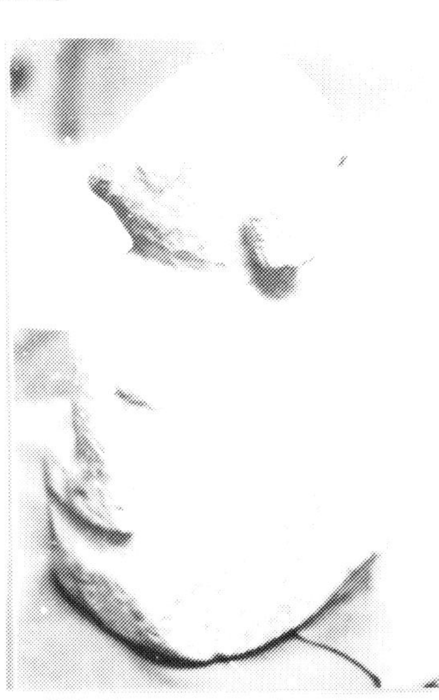

85. Reverse side of Right

PIPES

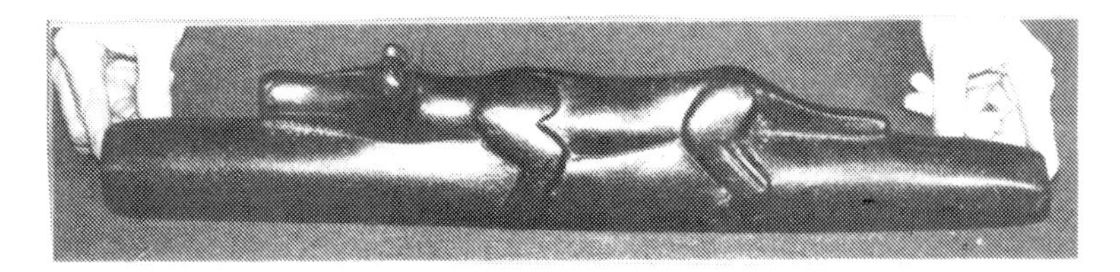

86. Blind Wolf Pipe
$100,000.00

87. Wood Duck Pipe
6-1/2" Steatite
$20,000.00

88. Wood Duck on a Tube Pipe
5", Rare

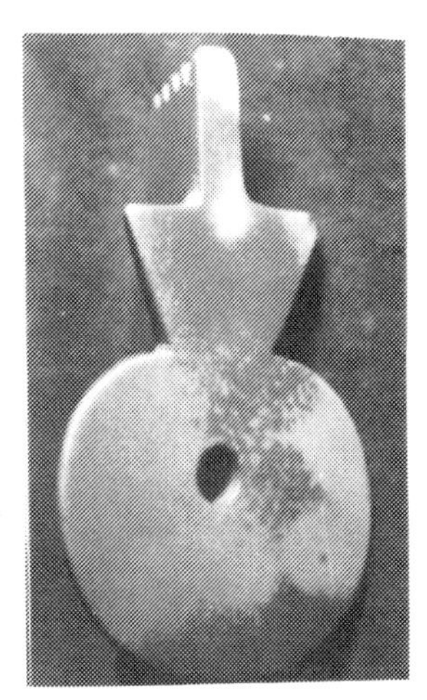

89. War Bundle Disk Pipe
Catlinite, 5"
$10,000.00

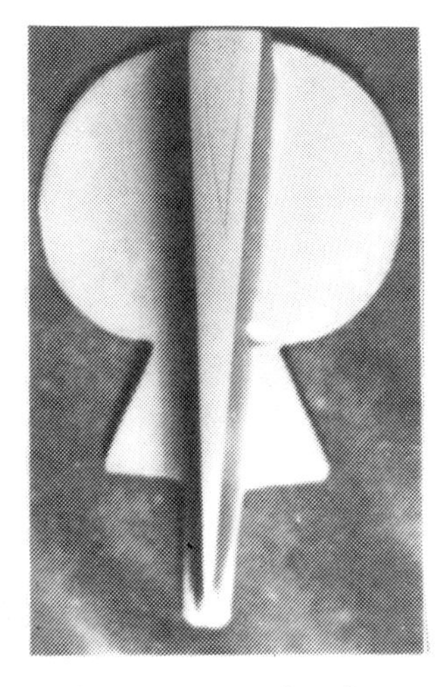

90. Reverse of Left

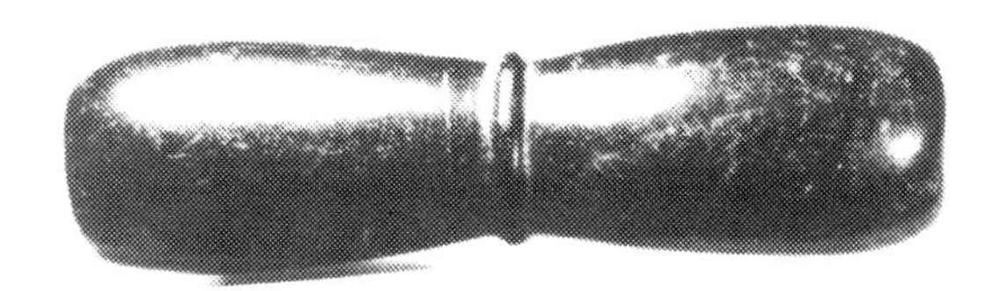

91. Medicine Tube, 10", Steatite
$5,000.00

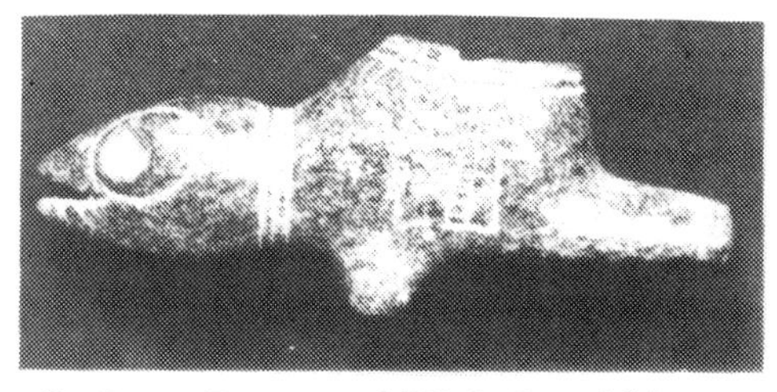

92. Turtle Pipe, Engraved With Spud-like Design, 3"
$1,500.00

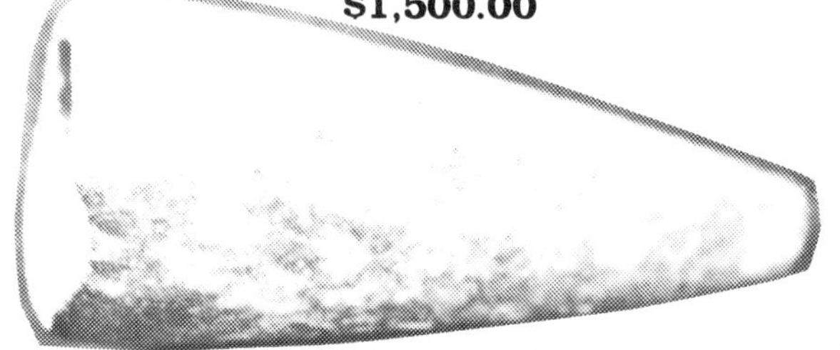

93. Archaic Tube Pipe, 6", Sandstone
$450.00

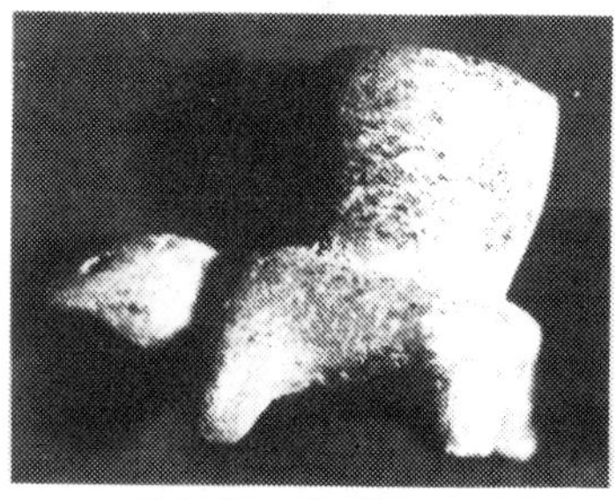

94. Turtle Pipe

$750.00

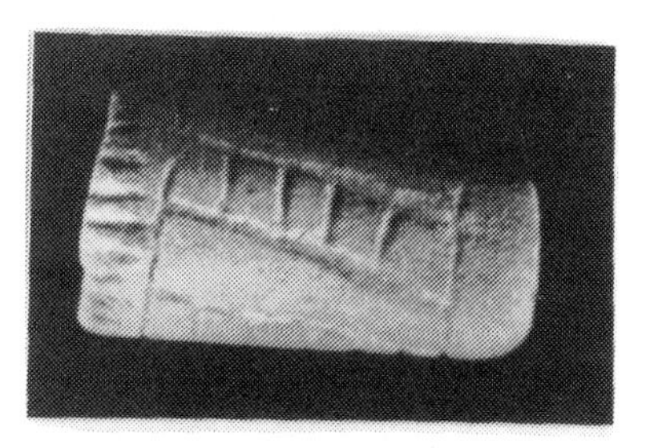

95. Miniature Archaic Tube, Engraved
$200.00

96. *Engraved Sandstone Elbow Pipe*
6", Some Damage
$800.00

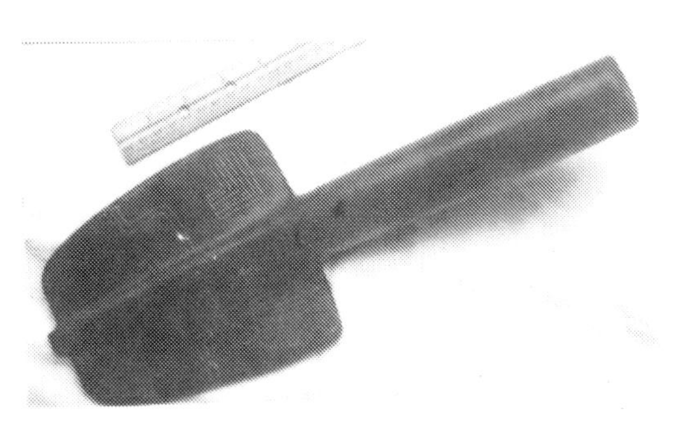

97. *Obtuse Angle Pipe*
Engraved, Repaired
$3,000.00

98. *Archaic Elbow Pipe*
Steatite, 5"
$750.00

99. *Engraved Archaic Elbow*
Pipe, 5"
$2,000.00

100. *Grasshopper Effigy, Restored, 5-1/2"*
$1,200.00

PLUMMET

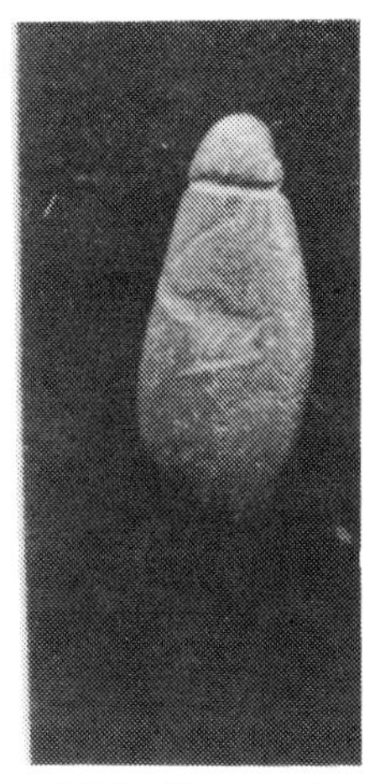

101. Greenstone Miniature Type

$50.00

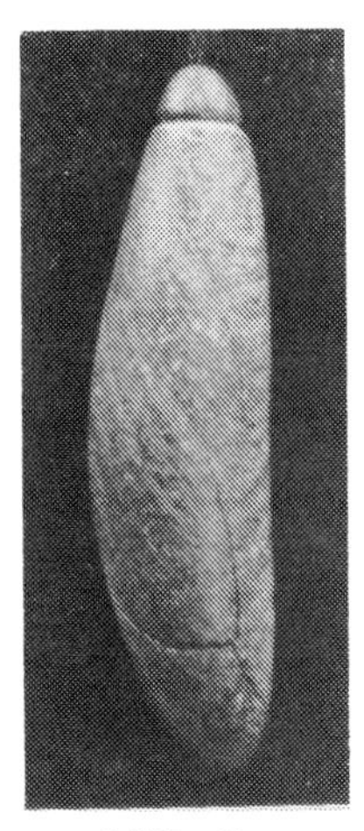

102. Green Slate

$100.00

103. Miniature Quartz Crystal Rare!

$500.00

104. Engraved Plummet

$450.00

105. Engraved Plummet

$550.00

DISCOIDAL (GAME STONE)

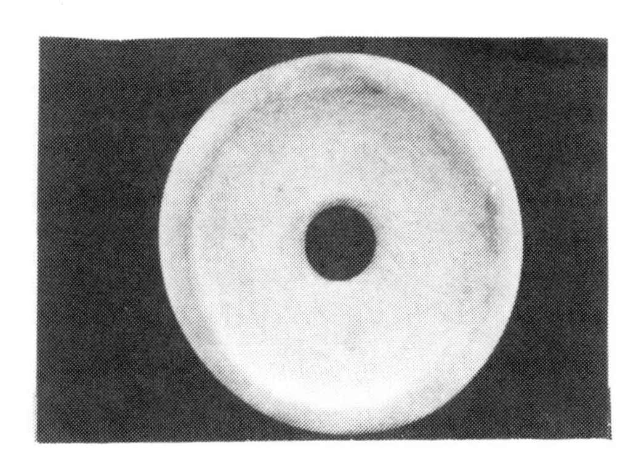

106. Perforated Type, 6" Diameter, Quartz
$2,500.00

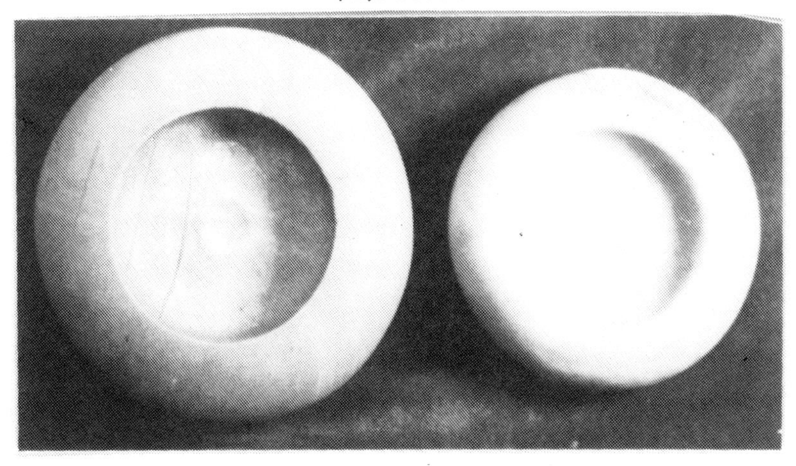

107. Tennessee Type Also Called Dimpled (note dimple in center) 6" Diameter
$3,500.00

108. Tennessee Type, Quartz 4" Diameter
$1,800.00

109. Double Cupped, Quartz, 4-1/2"

$1,100.00

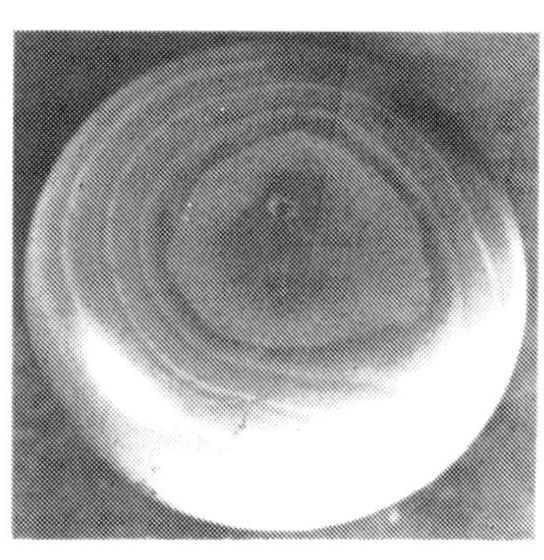

110. Biscuit Type, 4" Diameter, unusual example made of Hornstone chert

$750.00

111. Biscuit Type, 3" Diameter, Limestone

$150.00

PAINT PALETTE (SUN DISC)

112. Warrior Priest, Broken
$7,500.00

113. Sun Effigy, 6" Diameter

$4,500.00

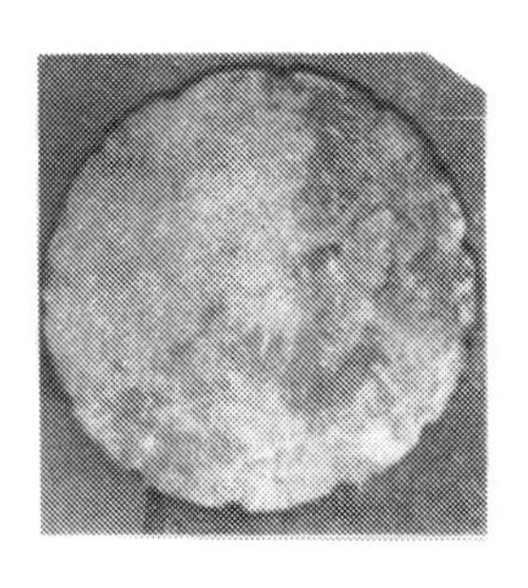

114. Scalloped Edge, 5" Diameter
$1,500.00

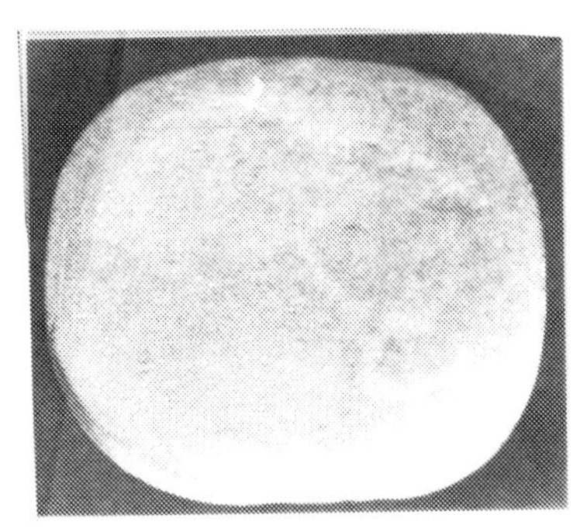

115. Rectangulated Type, 5"
$500.00

CHIPPED STONE

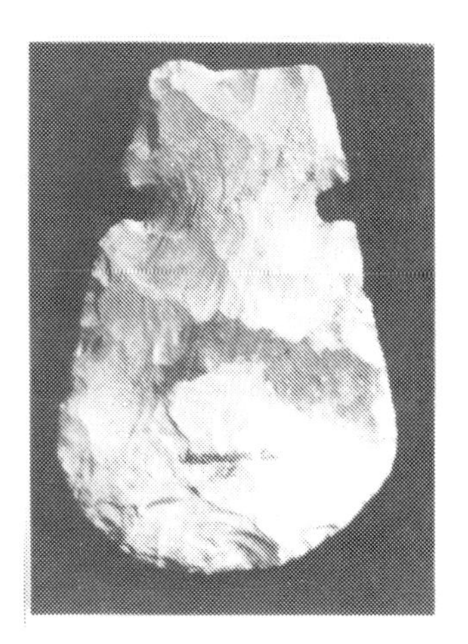

2 - *Notched Hoe, Approx. 5″*
$275.00

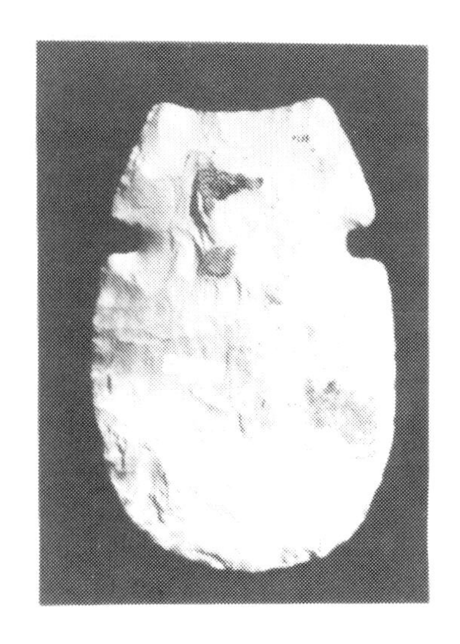

1 - *Notched Hoe, Approx. 6″*
$750.00

3 4 5 6 7

3 - *Dover Spade, 10″*..**$375.00**
4 - *Dover Spade, 11″*..**$500.00**
5 - *Dover Spade, 9″*..**$200.00**
6 - *Dover Chisel*...**$375.00**
7 - *Dover Chisel*...**$175.00**

8 - Dover Spade, 16"
$2,500.00

9 - Duck River Sword, 12" ... **$1,800.00**

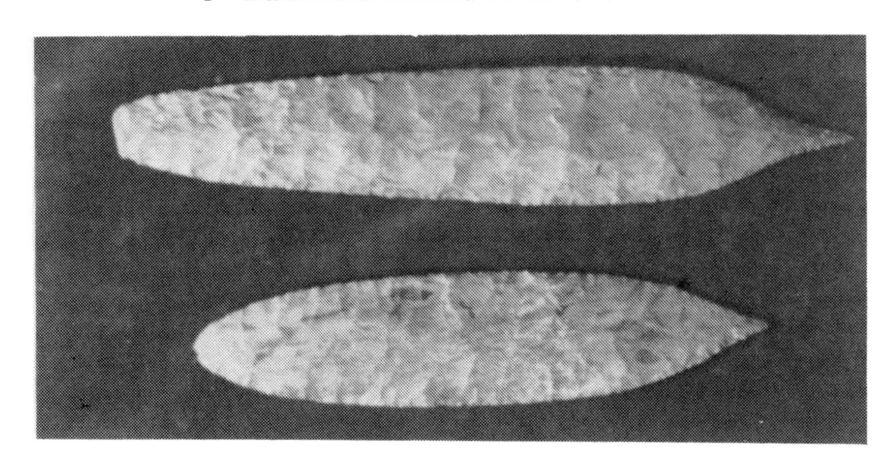

10 - Duck River Sword, 8-1/2" ... **$1,100.00**

ROCK DRAWING

Spider on Web
Approximately 14" Diameter
$3,500.00

ROCK DRAWING

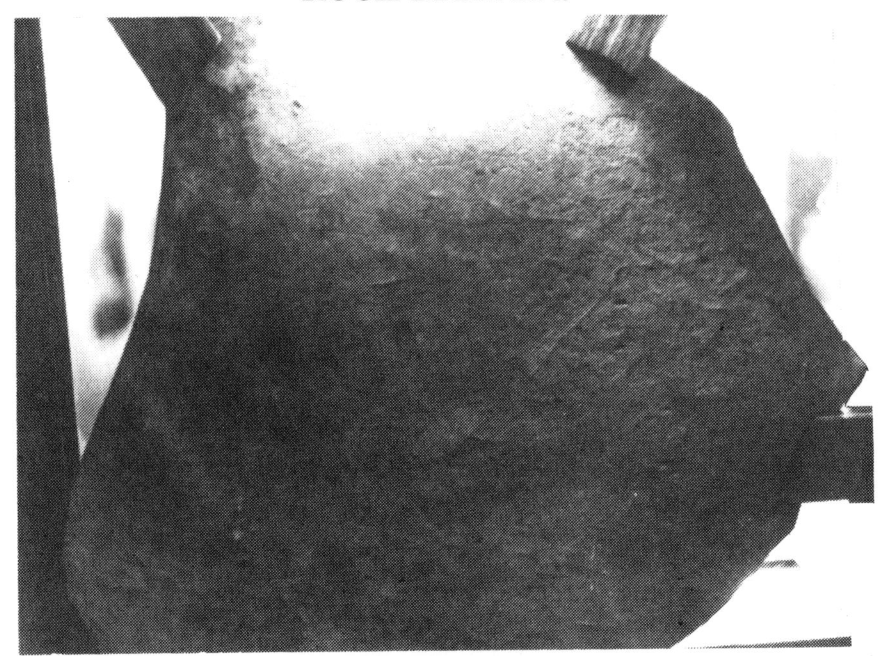

Double Eagle Dancer
Approximately 14" Diameter
$7,500.00

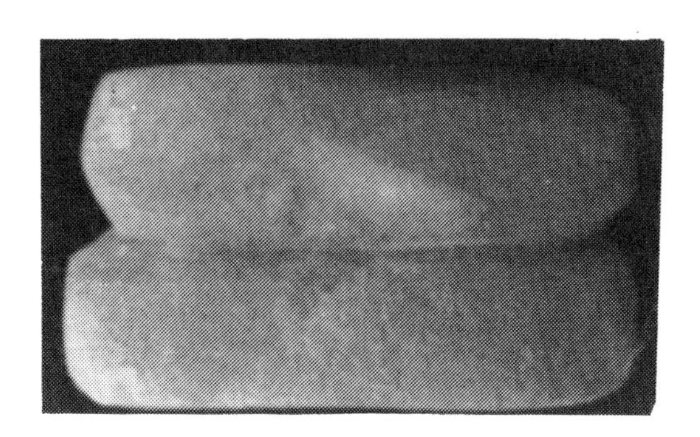

Net Sinkers
2" ... $25.00
3" ... $45.00
4" ... $65.00
5" ... $100.00

Flint Blades and Projectile Points of the North American Indian, by Lawrence N. Tulley
Collector Books, Paducah, KY - 1986

Field Guide to Point Types of the Tennessee River Basin, by Doug Puckett,
Custom Productions, Savannah, TN - 1983

Field Guide to Point Types of the State of Florida, by Son Anderson and Doug Puckett
Printed by *Fundco Printers*, Savannah, TN - 1984

A Guide to the Identification of Florida Projectile Points, by Ripley Bullen
Published by *Kendall Books* - 1975

Outlines and Other Data on West Central Florida Projectile Points, Compiled by George D. Robinson,
Published by *Central Gulf Coast Archaeological Society*

A Revised Edition of A Guide to the Identification of Florida Projectile Points,
Kendall Books - 1975

Handbook of Alabama Archaeology by Cambron and Hulse, Published by *The Archaeological Research Association of Alabama, Inc.* - 1975

Arrowheads and Projectile Points, by Lar Hothem
Collector Books, Paducah, KY - 1983

Eva-an Archaic Site by T.M.N. Lewis and Madelin Kneberg Lewis, *University of Tennessee Press*
Knoxville, TN - 1961

A Field Guide to Southesastern Point Types by James W. Cambron, Decatur AL

Indian Artifacts, by Virgil L. Russell and Mrs. Russell, *Johnson Publishing Co.*, Boulder, CO - 1962

Sun Circles and Human Hands, Emma Lila Fundeburk and Mary Douglas Foreman Published by authors, *Peragon Press*, Mongomery, AL - 1957

H. J. Holland - Personal Communication Leighton, Alabama - 1962

An Archaeological Survey of the Pickwick Basin the Adjacent Portions of the States of Alabama, Mississippi and Tennessee, by William S. Webb, Bureau of American Ethnology, Bulletin 129, Washington - 1942

An Archaeological Survey of Wheeler Basin on the Tennessee River in Northern Alabama, by William S. Webb and Charles G. Wilder, Smithsonian Institution, Bureau of American Ehtnology, Bulletin 122, Washington

An Archaeological Survey of Guntersville Basin on the Tennessee River in Northern Alabama, by William S. Webb, *University of Kentucky Press*, Lexington

Points and Blades of the Coastal Plain, by John Powell, Published by *American Systems of the Carolinas, Inc.*, West Columbia, South Carolina - 1990

American Indian Point Types of North Florida, South Alabama and South Georgia, Son Anderson - 1987

Guide to the Identification of Certain American Indian Projectile Points, Okalahoma Anthropological Society, Special Bulletin No. 1, Robert E. Bell - 1958

Guide to the Identification of Certain American Indian Projectile Points, Oklahoma Anthropological Society, Special Bulletin No. 2, Robert E. Bell - 1960

Guide to the Identification of Certain American Indian Projectile Points, Special Bulletin No. 3, Oklahoma Anthropological Society, Gregory Perino - 1968

Guide to the Identification of Certain American Indian Projectile Points, Special Bulletin No. 4, Oklahoma Anthropological Society, Gregory Perino - 1971

Selected Preforms, Points, and Knives of the North American Indians, Vol. 1, *Points and Barbs Press*, Gregory Perino - 1985

Central States Archaeological Journal, Numerous Volumes

Prehistoric Indians of the Southeast, Archaeology of Alabama and the Middle South, University of Alabama Press, John A. Walthall - 1980

Overstreets Indian Projectile Point Price Guide, No. 1, Published by Overstreet Publication, Inc., Cleveland, Tennessee, by Robert M. Overstreet and Howard Peake - 1989

The Indians of the Southeastern United States, Smithsonian Institution Bureau of American Ethnology, Bulletin 137, by John R. Swanton, Washington - 1946

Ten Years of the Tennessee Archaeologist Selected Subjects, Vol. 11, Published by the Tennessee Archaeological Society

Handbook of American Indians North of Mexico, by Smithsonian Institution Bureau of American Ethnology, Bulletin 30, Part 1 and 2, Edited by Fredrick Webb Hodges, Washinton - 1912

Handbook of Texas Archeology Type Descriptions, Published by The Texas Archeological Society, Special Publication Number One and the Texas Memorial Museum, Bulletin Number Four, Edited by Dee Ann Suhm and Edward B. Jelks, Austin, Texas - 1962

GLOSSARY

This glossary was compiled from words used in the text, and from A Glossary of Flint Flaking Terms in "A Comparative Study of Some Unfinished Fluted Points and Channel Flakes from the Tennessee Valley" (Cambron and Hulse, 1961), "Principles of Stratigraphy" (Grabau, 1960), and a glossary of archaeological terms by T. M. N. Lewis (1958). Words defined by Lewis are followed by the initial "L" and those by Grabau by the initial "G". Words not marked were defined by Cambron in Handbook of Alabama Archaeology, 1964.

Aborigine (L) - a native inhabitant of a country; in America, the Indian.

Archaic (L) - The culture which followed the Paleo-Indian period and preceded the early Woodland culture in the eastern United States. This was a pre-agricultural, pre-pottery culture. Food was obtained by hunting and gathering. Earliest date (Rocha-Carbon) in Tennessee is about 5000 B.C.

Arrowhead, Arrowpoint (L) - a weapon point generally regarded to be less than 2-1/2 inches in length. Longer points are regarded as spearpoints and knives.

Artifact (L) - an object of human workmanship, especially one of prehistoric origin.

Assemblage - a group of artifacts representing a culture.

Atlatl (L) - the Aztec word for spear-thrower. The device is a wooden stick with a hand grip at one end and a spur or hook at the other which fits into the socketed end of a spear shaft. The device lengthens the throwing arm and gives greater force to the spear.

Auricle - applied to the ear-like parts of a projectile point.

Auriculate - having auricles or ear-like parts.

Autotype (G) - a specimen not belonging to the primary material but identified with an already described and named type and selected by the nomenclator himself for the purpose of further illustrating his type.

Barb - a barbed shoulder of a point.

Base - the proximal end or area of a point nearest the haft. In describing point types, the base is assumed to be the lowest part of the point, thus the distal end is the uppermost part.

Basal Constriction - See Hafting Constriction.

Basal Grinding (L) - the grinding away of the sharp basal edges and lateral edges near the base of a projectile point to prevent cutting of the lashings.

Basal Thinning (L) - produced to remove small longitudinal flakes from the basal edge of a projectile point.

Baton Flaking - removal of flakes from stone by striking blows with a baton-like tool. A method of direct percussion flaking.

Bifaced (L) - a term applied to flaked stone artifacts that have been worked on both faces.

Blade - that part of a projectile point above the hafting area.

Blade Scar - a uniface scar surface resulting from a blade having been struck from a core. As another blade is struck from the core, it bears the blade scar of the previous blade.

Bulb of Percussion - a bulb resulting from a conchoidal fracture of stone.

Bulbar Depression - the depression left from the bulb of percussion when a blade or flake is struck from a core.

Ceramic (L) - pertaining to pottery.

Chalcedony - a flint-like stone with a waxy appearance. Alabama sources are usually nodular.

Channel Flake (L) - a long flake removed longitudinally from the face of a blade to form a flute or channel.

Chert (L) - an impure variety of flint.

Chip - See Flake.

Chronology (L) - the arrangement of events, or the material representing them, in the order of their occurrence in time.

Conchoidal (L) - shaped like half of a clam shell; refers to the characteristic fractures resulting from pressure and percussion flaking of flint.

Conglomerate (L) - conglomerate rock is composed of rounded pebbles and sand cemented together into solid rock.

Conoidal Theory of Flint Fracture - the theory that ideally a cone will be punched out of a piece of flint when it is struck with sufficient force.

Core (L) - a stone, usually flint, from which flakes have been removed by percussion.

Cotype (G) - an example of the original series when there is no holotype, the describer having used a number of examples as of equal value.

Culture (L) - the way of life of a group of people, comprising all their activities and beliefs. Archaeologically, a culture is represented by the material remains left by a group.

Cultural Complex (L) - a group of traits whose associations in time and space indicate that they were the products of the activities of a specific human group.

Deposit (L) - any accumulation laid down by human occupational activities.

Direct Percussion - flaking flint by striking it directly with a hammerstone or other object.

Distal (L) - when applied to a bone, it is the end farthest from the body.

Distal End - when applied to a projectile point or other hafted artifact, it is the end farthest from the point of attachment.

Drift - an implement, usually made of antler, used in indirect percussion flaking.

Ear - See Auricle.

Early Archaic - an early phase of the Archaic culture that may be represented by notched projectile points and is considered pre-shellmound in North Alabama.

Face - the area of a projectile point or tool between the edges. This may include the blade and hafting area.

Flake - in flint work it is a thin piece split from the parent material or core; to remove flakes from parent materials in flint work.

Flake Scar - a scar on the parent material resulting from the removal of a flake.

Flaking - the removing of flakes from a core or artifact in flint working.

Flaking Tool - an implement used in flaking stone tools; it is often made from an antler tine.

Flat Flaking - See Shallow Flaking.

Flute or Flute Scar - the scar left on the face of a projectile point as a result of fluting or removal of a channel flake.

Ground - areas, especially hafting area edges, that have been abraded smooth.

Haft (L) - a handle (or shaft); to provide with a handle.

Hafting Area - the area of a point or tool that receives the lashings, etc. in hafting procedures; the proximal end of a projectile point.

Hafting Constriction - a hafting area that is constricted along the side edges.

Hammerstone (L) - a hard pebble showing battering from use as a hammer.

Hinge Fracture - when the terminal end of a blade or flake, being struck from the parent material, makes a sharp dip into the material causing a deep fracture, it is called a hinge fracture. If a fracture of this type occurs during the removal of a channel flake from a projectile point it usually breaks at the terminal end of the channel flake.

Holotype (G) - among the primary types a holotype is the original specimen selected as the type, and from which the original description (protolog), or the original illustration (protograph), is made.

Homotype (G) - a homotype is a specimen not used in the literature but identified by a specialist, after comparing with the holotype.

Impact Fracture - a fracture of a projectile point resulting from impact during use; a shattering of the distal end of a projectile point as a result of impact with a resistant object.

Indirect Percussion - flaking flint by striking a drift which has been placed against the flint.

In Situ (L) - in place.

Lanceolate shaped like the head of a lance; of leaves, etc., narrow, and tapered toward the apex, or (sometimes) toward each end.

Lithic (L) - pertaining to stone.

Main Flute - the central flute struck from the face of a projectile point. This is usually the longest flute occurring during multiple fluting of a projectile point.

Median Ridge - a ridge left along the center of the blade of a projectile point during manufacture.

Midden (L) - the deposit of refuse generally present on a village site.

Mississippian Culture (L) - the culture that appeared in the Southeast around 1000 years ago. Shows strong Mexican influences and is associated with many groups ancestral to the historic Muskhogean speaking tribes of the Southeast.

Multiple Fluting - multiple fluting is the result of the removal of more than one channel flake from one face of a projectile point. This is usually accomplished by two primary flutes and a main flute being made.

Obsidian - a volcanic glass; may be black, brown or green in color.

Obverse Face - as a means of identifying the faces of projectile points, the obverse face is the one without marks, such as site number, etc.

Paleo-Indian (L) - a name assigned to nomadic groups who were the first inhabitants of the new world. Their culture was comparable to that of the late Paleolithic of the old world.

Paleolithic (L) - a term applied to the Stone Age of the old world. During this period man had no knowledge of plant and aminal domestication and no knowledge of pottery and metals.

Patina (L) - an adhesive crust or discoloration produced by weathering of an object. It does not necessarily imply great age.

Percussion Flaking (L) - removal of flakes from stone by striking blows with a stone or other hammer.

Pitch of Striking Platform - the degree of angle resulting from the removal of flakes from the base of a point to make a striking platform.

Plano-convex (L) - a term used to describe an object that is flat on one face and convex on the other.

Plesiotype (G) - a plesiotype is a specimen not belonging to the primary material but identified with an already described and named type and selected by someone else than the original describer for the purpose of further illustrating the type.

Pressure Flaking (L) - shaping a stone such as an arrowpoint by removing flakes from the edges by pressure with a pointed implement made of material such as bone or antler.

Primary Flaking - the first rough series of flakes removed in shaping blade or hafting area edges of projectile points or tools.

Primary Flute - usually one of two short flutes removed from the basal face of a point in order to leave a striking nipple.

Projectile Point (L) - a pointed artifact used on a spear, arrow or dart.

Proximal End - the end of a projectile point to which a haft is attached. This is part of the hafting area.

Quartz (L) - a material frequently used in projectile points and other artifacts. When quartz is clear and colorless, it is called rock crystal; milky quartz is milky white; smoky quartz is a cloudy brown color; rose quartz is a pale red color; sugar quartz is the color of brown sugar.

Quartzite - a granular form of quartz, often quartz fragments cemented together.

Resolved Flaking (L) - the method of striking flakes from a flint core by directing the blow inward. A resolved flake struck off in this manner is thinner and narrower at the percussion end; the flake is thicker and wider at the opposite end, and, when it does not extend the full length of the core, it usually ends in a hinge fracture.

Retouch (L) - a term applied to the secondary removal by pressure of small flakes from the edge of a flaked stone artifact to produce sharpness.

Reverse Face - the reverse face of a point may be designated as the one with the site number, etc., marked on it. This is assumed to be the less finished side.

Reworked - a projectile point or other artifact is said to be reworked when the shape has been altered by flaking, grinding, etc., either by the manufacturer or other people.

Rind - a deeply weathered area on the outer surface of a nodule or chunk of flint or like material.

Rounded - a term applied to a relatively symmetrical curved area of a projectile point.

Secondary Flaking - the removal of small flakes, usually by pressure flaking, with a piece of bone or antler. The secondary flaking along the edge of the blade of a point is usually designed to finish the blade edges.

Serrated - having intentional toothed projections along an edge.

Shallow Flaking - the removal of shallow flakes in shaping a projectile point or tool.

Shellmound Archaic - that part of the Archaic period when shellfish middens were formed as residue of the gathering economy of the people in an area.

Spall (L) - waste flake struck from a larger piece of flint.

Spear Thrower (L) - See Atlatl.

Stem - a type of hafting area of projectile points.

Stratification (L) - formation in strata or layers. When village site deposits show more than one stratum formed by successive occupations by groups of people, the lowest stratum is the oldest.

Stratigraphy (L) - the arrangement of strata with respect to position in which they were laid down by human occupation or from natural causes.

Striking Nipple - a nipple left near the center of the striking platform where a drift may be set to strike the main flute. Usually the nipple is formed by the removal of two primary flutes.

Striking Platform - a prepared basal edge of a projectile point. This edge is beveled to a degree of pitch that will allow a drift to be set at the proper angle to strike off a channel flake.

Trait, Culture Trait (L) - any object or other evidence that is the result of human behavior or action.

Transitional - of, pertaining to, characterized by, or involving transition; intermediate. In archaeology, usually pertaining to evidence of transition from one culture to another.

Transitional Paleo-Indian - a cultural period intermediate between Paleo-Indian and Archaic.

Typology (L) - a study of arrangement of specimens separated into types.

Unifaced (L) - a term applied to flaked stone artifacts that have been worked on one face only.

Variant - in projectile point typology a variation of the type described as of the original series.

Vein Quartz - a relatively pure type of quartz found in veins in igneous areas.

Woodland Culture (L) - a widespread culture in eastern America which appeared in the Southeast about 2000 B. C. The presence of pottery differentiates it from the Archaic which preceded it.

Worked - in projectile point descriptions, describes an area of an artifact that has been shaped or altered by man - - such as the removal of flakes along a blade edge.